TIME'S PENDULUM

The Quest to Capture Time—
From Sundials to Atomic Clocks

TIME'S PENDULUM

The Quest to Capture Time—
From Sundials to Atomic Clocks

Jo Ellen Barnett

PLENUM TRADE • NEW YORK and LONDON

Library of Congress Cataloging-in-Publication Data

On file

ISBN 0-306-45787-3

© 1998 Jo Ellen Barnett
Plenum Press is a Division of Plenum Publishing Corporation
233 Spring Street, New York, N.Y. 10013

http://www.plenum.com

Printed in the United States of America

To the memory of Geren (1971–1979),
who didn't get enough time

It was propped against the collar box and I lay listening to it. Hearing it, that is. I don't suppose anybody ever deliberately listens to a watch or a clock. You don't have to. You can be oblivious to the sound for a long while, then in a second of ticking it can create in the mind unbroken the long diminishing parade of time you didn't hear.

—William Faulkner
The Sound and the Fury

Contents

Part One

The Time of Day

Introduction

The year 2000 A.D. is upon us. It will be no time at all until the ball falls in Times Square and we cheer as the last moments of the second millennium A.D. escape us and we enter January 1, 2000. But those moments, in terms of time itself, will be no different from the moment I am writing this or you are reading it. Time, that intangible element as dear to us as the air we breathe, knows nothing of our demarcations of it. It existed long before we were here to ponder it or cut it up into little pieces, and it will exist long after we are gone. It is one of the primal elements of our lives.

Anyone today could agree with St. Augustine, writing at the turn of the 5th century A.D.:

"For what is time? Who can readily and briefly explain this? Who can even in thought comprehend it, so as to utter a word about it? But what in discourse do we mention more familiarly and knowingly, than time? And, we understand, when we speak of it; we understand also, when we hear it spoken of by another. What then is time? If no one asks me, I know: if I wish to explain it to one that asketh, I know not . . ."[1]

However long the human race may survive, it is hard to imagine that the time will ever come when St. Augustine's statement will seem in any way dated. Like St. Augustine, probably no human being ever gets through life without many moments of trying, but failing, to grasp this gossamer and yet so terribly real phenomenon.

The story of time is not like that of, say, matter, about which new concrete entities are being discovered seemingly everyday. Ever since the elements began to be discovered in the eighteenth and nineteenth centuries, there has been a steady and spectacular unraveling of the intricacies of matter. John Dalton discovered the atom in 1805, J.J. Thomson and Ernest Rutherford opened up the subatomic realm of electrons and protons a century later, and in the past few decades other minuscule particles like quarks and neutrinos and muons have been found. At the opposite end of the size spectrum, new configurations of matter like pulsars and quasars and black holes have been discovered. Along the way, molecules, radioactivity, atomic fusion and fission, DNA, viruses, and the chemical systems of life are just a few of the endless facts about matter which have been revealed. Nothing of this sort is happening with time. There are no names for things which make up the structure of time: no timelets, or subtimelets, or macrotimes. It's all still just the one entity—time.

Granted, stupendous discoveries have been made during this century and the last concerning time. The mapping of the unfathomable vastness of the ages of the earth and universe has changed forever our notion of time and, with it, our conception of the world and our place in it. At the beginning of this century, Einstein's theory of relativity replaced Newton's concept of absolute time with that of relative time. For over two hundred years, Newton's view of time as an unalterable phenomenon independent of everything external to it held the day, and this is still how we regard time in our day-to-day lives on the surface of the earth. But we know now that in the broader universe where distances are measured by the speed of light, time lacks the sovereign integrity which Newton gave to it. We are still struggling to pull ourselves away from the old notion and grasp a time which is inextricably bound to space and which depends upon the position of the observer. From mid-century, the debate between the Big Bang and Steady State theories of the origin of the universe gave us powerful reverberations of the ancient dichotomy between linear and cyclical time. The apparent outcome in favor of the Big Bang theory has put us up against the wall with questions like: Did time have a beginning? If so, will it have an end? When did the former take place and when might the latter occur?

These discoveries change the *extent* of time and its *nature*, but it still remains the one, uniform phenomenon. It seems that time has no structure, but only a character which can be universally applied to it. In comparison with the intricate and exotic world of matter, time looks quite colorless. What is there to say except that there is an enormous amount of it, that it is relative rather than absolute, and that, incomprehensible as it might seem, it appears that it not only had a beginning but that it may have an end as well?

We can begin to answer that question only by looking in another direction—away from the raw phenomenon of time and toward ourselves, toward what we have done to give it the structure it does not have in itself. We have divided time into segments of every conceivable size: nanoseconds, milliseconds, seconds, minutes, hours, days, weeks, months, years, decades, centuries, millenia, eons, eras and on and on. But these divisions differ fundamentally from the building blocks of matter. Only the day, month, and year are based on natural phenomena; all of the rest are man-made divisions, and none of them is intrinsic to time as are the structural particles of matter. They are not discoveries *about* time so much as inventions of the human mind for the purpose of carving order into something that cannot be seen, or heard, or felt.

What are the thickest threads in the tapestry of time that we have woven about ourselves? The oldest is the calendar, which orders the time it takes for the earth to make one complete circuit of the sun. The years in their turn have been strung together to make the chronologies which help us hold onto our past. Although the Christian calendar and system of chronological reckoning have now become the worldwide lingua franca for public affairs and commerce, the rich variety of possible means of configuring the earth's circuit of the sun is only hinted at in the enduring calendars of the Jews, Moslems, and Chinese. Calendars are among the most stable and long-lived of all our systems for ordering time.

Our other two great systems for counting time—which this book will explore—lie to either side of the calendar and differ as much from each other as they do from it: the mechanical clock and the radioactive "clocks." The clock counts the tiny bit of time encompassed by one rotation of our earth upon its own axis, and our instruments for measuring the half-lives of radioactive atoms count the inconceivable eons of the earth's lifetime. The domain of the clock is a grain of sand on the endless beach of the earth's whole life.

Not only do the time periods encompassed lie at opposite ends of the scale, but the effects on our lives of these two great time counters could not be more at odds. The clock reaches deep into the interstices of our everyday lives, while the radioactive clocks simply offer for contemplation facts about our planet more awesome than we can grasp. Their histories are different, too. Humankind has lived with some form of the clock for nearly four thousand years, while the radioactive clocks burst upon us less than a century ago.

The traditional clock and the radioactive clocks count different lengths of time and they count them for different purposes, and in the process they also reveal the Janus-faced aspect of time. On the one hand, all that we can ever have of time is the evanescent present moment. The clock is the strange instrument we have invented to record the present even as it flees. On the other hand, we have built a world upon such transcience, and if we can never bring the past back, we can in some sense keep hold of it through remembrance. Just as the clock is oblivious of the past, so the radioactive clocks know nothing of the present. Instead, their power to "remember" a past too remote for our wildest imaginings has brought to life a history of our planet which we could otherwise never have known.

What can two such different instruments for time-telling, measuring off such staggeringly divergent realms of time, have in common with each other? They share with the calendar the quintessential requirement of all timekeepers: the segment of time they measure is constant. It does not matter how long it is; it matters only that it never changes. A minute is a minute is a minute. A year is a year is a year. A particular radioactive atom's half-life remains invariant though eons glide by under its gaze.

Such unvarying time segments are extremely difficult to create mechanically, and they are equally hard to find in nature. Because our objective knowledge of time could not exist without them, the history of time-telling has been largely a search for more and more constant temporal segments. As we will see, it has been nature, and not human invention, which has been responsible for the best of them.

The clock that tells us the time of day and the radioactive clocks which chart the earth's history meet at two other points, neither of which is shared by the calendar. First, they both require scientific instruments to do their measurements. Second, in today's world the final arbiter of both their time segments is the *atom*. When we call to get the time, it is 9,192,631,770 oscillations of the radiation emitted when the outer electron in the cesium-133 atom flips over that is ultimately responsible for the answer we hear. Likewise, when we read that the earth formed 4.5 billion years ago, or that our *Australopithecus afarensis* ancestor "Lucy" lived 3.18 million years ago, or that seeds from a cultivated squash plant found in a buried Mexican village are ten thousand years old, it is the constant decay rate of radioactive atoms that has supplied the proof. If time itself is mysterious, it has something of a rival in the surreal bond which the atom has made between the shortest and longest of the threads we have woven into our tapestry of time.

In addition to this bond and to their own intrinsic interest, these two "timepieces" are the subject of the present book because of a third point of commonality: unlike the case of the clear and simple calendar, most of us don't really understand them. Consider the clock. We ask: "What time is it?" We look at our watches and answer 9:43 P.M., or 7:05 A.M., or whatever, and that's the end of it. We've answered as completely as we know how, and we're ready to move on to something more interesting. Not only does our answer reduce time to "clock time," but it unthinkingly skims the surface of an intellectual and cultural heritage whose roots go at least as deep as ancient Egypt and Mesopotamia.

The ubiquity of our clocks and watches, and the time-obsessed culture of which they are both cause and effect, has led us to regard our way of timekeeping as coming with the territory, like the air we breathe. Nothing could be further from the truth. The watches on our wrists today are the culmination of a long, slow, arduous process of invention and discovery which ultimately began about four millennia ago when someone stuck a stick in the ground and watched its shadow change throughout the day. Like the sundial and its other forebears, our clock is a scientific instrument which is a repository not only of our mechanical ingenuity but of central facts about the physics of our planet. In Part I, we will try to reconnect our clock time to those natural facts as well as to the cultural situations which brought about the need for their discovery.

In Part II we will leave the present and look at the ideas about the past—and the nature of time itself—that people have come up with throughout the ages. In order to gain a better understanding of how divergent these views were, we will look at how each answered the question: "How old is the earth?" It was not until late in the seventeenth century that empirical evidence made its first real chink in the armor of fixed ideas about that answer. It required two more centuries for a fact-based story to slowly take the place of received beliefs. Finally, at the turn of the twentieth century, the natural, constant rate of decay of radioactive atoms was discovered, and empirical knowledge of the earth's history could at last be quantitated in terms of time. At least for the earth, we have answered the question of *how much* time, but the cloud of unknowing about *what* time is seeps out from our hard core of fact as incessantly as it did for our forebears.

Without being able to say what time is, we have knocked ourselves out learning how to measure it. Our efforts to measure time have taught us many things, but they have not brought us through to the phenomenon itself. They have increased our knowledge of the physical world and have held up a mirror to our own unique culture, but time itself . . . We have cast our nets to find time and have only come up with other things.

Chapter 1

The Planetary Basis of Our Day

Time is forever dividing itself into innumerable futures . . .
Jorge Luis Borges

Let's begin at the beginning. Before delving into the genealogy of our clock time, we need to take a moment to look at the natural "given" upon which it is based, and realize that it came about quite by chance. Of the nine planets which circle our sun, six have rotation rates between about ten and twenty-five hours: Mars is the closest to us, with a rate of 24.6 hours, while Jupiter, Saturn, Uranus and Neptune have rates of 9.9, 10.7, 17.2, and 16.1, respectively. Pluto, the peculiar planet, takes 6.9 days. The real anomalies are the innermost planets: Mercury takes fifty-nine *days* to turn upon itself, and Venus actually has a day that is longer than its year: it circles the sun in 225 days, but requires 243 *days* to turn upon itself. Only twice in each Venusian year is there a sunrise and sunset. Did something happen to Mercury and Venus to jog them so far out of the 10–25 hour range, or is that range for the other six planets just a coincidence? Although most planetary scientists believe that something did happen to Mercury and Venus, they don't know what it was or how it happened.

We do know that if the accident of our own planet had occurred a little differently, we would have been dealt a very foreign world in terms of our cycle of night and day. Try to imagine life on a planet with a six-hour rotation period, or even one of ten, fifteen, or twenty hours. There's no point in trying to imagine life with a light/dark cycle like that on Mercury, because it probably wouldn't be viable for human beings or any form of life. Even if Mercury had an atmosphere we could breathe, any particular place on this planet endures six weeks of baking sunlight followed by six weeks of freezing darkness.

The quirk of our having a twenty-four-hour day instead of one of some other length is compounded when we realize that our twenty-four hours is just a temporary phenomenon. In fact, it is only we humans who have been dealt a twenty-four-hour day, *not* our planet. In order to explain this, we have to take a look at the moon. Our planet could have done without the moon, but whether we could have is another story. The moon is definitely not just decoration.

So crucial is the moon to our lives that it is actually incorrect to say that without the moon we wouldn't have a twenty-four-hour day, because *we* wouldn't be around to notice. Some other form of intelligent life might be, but it wouldn't be *us*. Without the moon, the earth would have evolved into a very different world than the one we know. Instead of taking twenty-four hours to rotate upon itself, it would now be taking only about *eight* hours, and each year would contain 1,095 of those eight-hour days. Without the moon, our tides would be only about a third as high, but the winds of this fast-turning planet would be far, far stronger. The atmosphere would have more oxygen, the magnetic field would be three times stronger, and, above all, the plant and animal life which would have evolved would be very different. Among other things, their biological clocks would be entrained to an eight-hour cycle and not to one of twenty-four hours.[1] The earth would be the same planet, but another world.

To compound things further, the event which gave us the moon and set into motion all of the changes it would create was, quite simply, an accident. It might easily not have happened, but as is the nature of accidents, it *did*.

The earth, like all of the other planets, formed between 4.5 and 4.6 billion years ago out of a massive nine-billion-mile-wide disk of dust particles which orbited the newly-created sun. As the particles collided, they sometimes bounced off each other, but just as often they bonded to form larger and larger particles. The most massive particle at the earth's distance from the sun finally attained sufficient gravitational force that it began to draw more and more particles toward itself. Eventually, it became our earth, while the same process in other regions of the solar system gave birth to the other planets.

The planets gathered up most of the debris in the primordial disk, but certainly not all of it. Thousands of planetesimals still orbit the sun as asteroids, and the several tons of meteors pelting our earth each year are reminders of that earliest world. Although we are unaware of it, almost every night we have before our eyes another reminder of that nascent time. At the same time that the planets were forming, a huge planetesimal about the size of Mars was also taking shape, and it followed an unusually elongated elliptical orbit. It is now believed that this giant object eventually struck the earth and caused the moon to be formed. If one wishes to tincture the facts with a teleological outlook and call it destiny, one may, but the event speaks just as loudly for an interpretation of pure chance.

Here is the theory about the moon's formation which prevails today: for millions of years, this gigantic planetesimal traveled trillions of miles through the solar system, sweeping past other planetesimals as well as by Mars and Venus. Each time it came close to another body, its orbit would be altered by the gravitational field of that body. Had it formed in an orbit that differed by just a few inches from the one it actually took, it would never have hit the earth. The encounters with the gravitational fields of other bodies would have stretched that few inches of difference into thousands of miles by the time the planetesimal approached the earth, and the collision we got would instead have been a miss. The close encounter with the earth would have again diverted its course, and it might have gone on to strike another planet someday or even to leave the solar system altogether.[2]

Fortunately for us, that is not what happened. All evidence today indicates that it *did* strike the earth, and more than five billion cubic miles of the earth's crust and mantle shot up into the air. At first, all this material orbited the earth just as, millions of years earlier, the disk of planetesimals had orbited the sun. Slowly, the debris accreted into larger bodies, until the gravitational force of the largest began to pull all the others toward it, and the moon materialized. For the moon, our planet is literally Mother Earth.

Of the other inner planets—Mercury, Venus, and Mars—only Mars has moons, but these two tiny bodies are believed to have been captured by Mars' gravitational field after they were already formed. The earth is thus unique among the solid inner planets in having a huge satellite whose diameter is more than a quarter that of its primary planet. In fact, if Pluto's captured moon Charon is excepted, our moon is *by far* the largest satellite in the solar system relative to the primary planet around which it orbits.

There are still many things we don't know about the moon, but after the Apollo moon missions of 1969–1972 we now have hard evidence about its age. Because the earth is such an active planet, it is difficult to find many rocks more than a billion years old, but a few have been found as old as 3.9 billion years. The moon, without atmosphere or water, is not subject to the forces of erosion that have destroyed so much of the evidence of earth's earliest state, and the Apollo astronauts were able to bring back many rock samples as old as four billion years, and some nearly as old as 4.5 billion years. We now know definitively that the moon has been around for a long, long time—in fact, within millions of years as long as the earth.

Although the moon may have been earth's satellite since almost the very beginning, their physical relationship has changed drastically during the intervening eons. In the past the moon was much, much closer to the earth than it is now. What does that mean for us? We were not around four billion years ago to see a giant moon looming in the sky, and because of its much greater gravitational pull, causing every tide to be a tsunami. We couldn't have breathed the CO_2-rich atmosphere anyway. What it means for us is that our twenty-four hours is only the present rotation period of the earth—it was much shorter in the past and will be longer in the future, and the reason for this has to do with the moon.

The six-hour rotation period mentioned earlier was not some sci-fi fantasy. In fact, many scientists believe this was the rotation period of the earth in its infancy.[3] Night and day switched on and off about every three hours on our violent, moonless early earth. How in the world did we get from this six-hour day to our own twenty-four-hour day? The answer to this question has everything to do with the moon, so let us first look more closely at where and how the moon entered the life of the earth.

Because of their mutual powers of gravitation, there is a limit to how close two rotating planetary bodies can be. Too close, and the force of gravity of the larger body prevents a smaller one from coalescing—its matter forever remains orbiting rubble which can never accrete into a solid body. In 1847, M. Edouard Roche put this limit into mathematical terms, known ever since as the Roche Limit. We know that our moon coalesced beyond the earth's Roche Limit—about seventy-three hundred miles from its surface—because otherwise our night sky would be constantly clouded by a ring of debris rather than lit by the bright phases of our moon. (The rings of Jupiter, Saturn, Uranus, and Neptune are all within their respective Roche Limits, while their moons all lie beyond this limit.)[4] It was pure chance that the Mars-sized object hit the earth, and again pure chance that the piece of earth it splintered off was propelled far enough to become our moon and not just orbiting debris. Such a ring of debris would have forever obscured our star-lit night sky, transformed our daylight sky, and vastly limited our chance to learn our whereabouts from the stars.*

We still don't know for certain how far away the moon was when it came together. Computer simulations of the collision indicate that our moon may have accreted about half again as far away as the Roche limit, or about eleven thousand miles above the earth's surface. Try to conceive of that surreal world: through the fierce winds generated by our earth's six-hour rotation, a giant moon would fill the sky,[6] so close that a viewer (if there had been any) could almost have counted the pebbles in its early craters. Another view holds that the moon

---

*Since the 1995 discovery of a planet orbiting a star fifty light years away in the constellation Pegasus, the scenario for the formation of planets just described may have to be revised. This planet has half the mass of Jupiter and yet appears to orbit its star at a distance of only five million miles—seven times closer than tiny Mercury orbits the sun. Most explanations of this "impossible planet" (and similar ones discovered since) indicate that it actually formed much farther out and subsequently migrated inward, but the questions it raises will not be resolved tomorrow.[5]

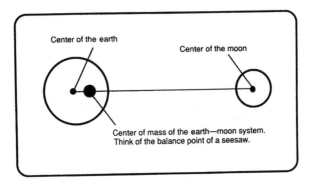

1. The earth–moon system.

may never have been more than about twice as close to the earth as it is today.[7] We may never know for certain, but we *do* know that in the past it was much closer, and that it has been slowly moving away ever since.

What is causing it to move away? "Strictly speaking, the moon does not revolve around the earth, but rather the earth–moon system swings around a common center of mass."[8] Because the earth has eighty-one times more mass than the moon, this center of rotation is buried over one thousand miles inside the earth. The moon and earth must remain in equilibrium, and they do this by obeying the law of conservation of angular momentum, which dictates that when the speed of motion decreases, the distance from the center must increase in compensation.[9] We have all seen this principle in action when watching a skater spin faster and faster as she pulls in her arms and then slower and slower if she brings them back out again. Her body is simply obeying the law of conservation of angular momentum.

Exactly the same principle applies to our earth–moon system. When the moon was much closer to the earth, the whole system spun much faster. But the moon's gravitational pull creates the ocean's tides, and friction occurs as water is dragged over shallow ocean floors. It also causes tides in the earth's solid crust, and more friction is created as layers of rock rub against each other. This friction dissipates some of the earth's energy of rotation by converting it into heat energy, so that the earth slows down. Because the total angular momentum must remain the same, as the earth slows down, the moon speeds up, and in doing so, it spirals slowly away from the earth. Let's return to the analogy of the skater. If we think of the earth as the skater's body and the moon as her hands, then the early earth–moon system was like the fast-spinning skater with her hands at her sides, while the system we have today resembles the more slowly revolving skater with her hands held out from her body.

Although to a far lesser extent, the sun's gravitational force also creates tides. This is why our highest tides occur when the sun and moon are lined up together. If the giant planetesimal had missed the earth and we had thus remained moonless, the tidal friction caused by the sun would by now have only slowed the earth to an eight-hour day rather than the six-hour period of rotation with which it began.

The far greater transformation of our day's length created by the earth–moon system has not followed a linear path throughout the eons. The *rate* at which the earth is slowing down is itself slowing down and has been ever since the earth–moon system first came into being. Since there was greater tidal friction when the moon was closer to the earth, much more of the earth's rotational energy was lost to heat, and it slowed down more quickly, causing the moon to move away at a far faster rate then than now. In fact, it has been estimated that by 4 billion years ago, the day had become about $13\frac{1}{2}$ hours long.[10] Thus, between then and now the moon's outward movement has slowed down enormously. Today it is only about two inches a year.

This rate has been quite precisely determined using an experiment set up by the astronauts of the Apollo 12 flight in 1969. The astronauts set up an array of reflectors on the moon, and laser beams were then shot at them from the earth; the beams bounced off and returned to earth. The distance to the moon was calculated simply (and very accurately) from the speed of light and the time the trip took. By repeating the experiment through the years, the two-inch-per-year rate of the moon's recession from us could be determined.[11]

We don't of course have hard evidence of the earth's shorter day during the earliest eons of its existence, but we have recently found evidence that around 900 million years ago, a day was only 18 hours and 10 minutes long! The rise and fall of tides leaves microscopic stripes in sedimentary rocks, which can be counted and translated into time intervals just as tree rings can.[12] This new evidence fits well with earlier calculations based on the bands of calcium carbonate which some types of coral lay down each year. These corals also deposit more calcium carbonate during the day than at night, so that very fine *daily* bands are also created. Paleontologists have found four hundred of these daily bands in coral dated to 400 million years ago, indicating that there were four hundred days in the year when primitive amphibians and the first forests were appearing.[13] Because the time required for the earth to circle the sun is changing so slightly that it can here be regarded as constant, these greater numbers of days in the year indicate that the earth was rotating faster back then and that the day was shorter. The coral ring data give a day of a little less than twenty-two hours 400 million years ago. During the time of the dinosaurs, it is believed that the earth had slowed to give a day of about $22^3/_4$ hours.[14] At present, during the reign of the twenty-four-hour day, only one second is being added to our day in 62,500 years.[15] Nevertheless, such is the power of time that in another 225 million years that would bring the earth to a twenty-five-hour day.

If we hadn't gotten a moon, we would have had nothing to cause the tides that created the friction that slowed the earth down, and our earth would be revolving nearly as fast as it was when it began. There is no reason that life, even intelligent life, could not have evolved on such a moonless, windy, fast-spinning planet, but it wouldn't have been us. A different kind of intelligent life than human beings might have evolved, and they would be as neatly adapted to their shorter day as we are to our twenty-four hours.

The point of all this is to give a picture of the bedrock upon which our twenty-four hours is based, and to show that it is not bedrock at all, but a slowly changing physical system that came into being by chance when a giant planetesimal slammed into the earth and gouged out the material which became our moon.

Chapter II

The Sundial and Its Temporary Hours

We live in an old chaos of the sun
Or old dependency of day and night, . . .
Wallace Stevens
Sunday Morning

Our ancestors, of course, had no idea of any of the things discussed in the previous chapter. They were at least as ignorant of our twenty-four *equal* hours—without which we can almost not talk (or think) about our day—as they were about any of our astronomical discoveries. Unlike these discoveries, our hours are imaginary divisions of time that are real only in our minds. Except for astronomers, no one conceived of dividing the day into twenty-four equal hours until almost three millennia after the first crude instrument was made for partitioning the day's time. Try for a moment to erase from your mind these equal hours and ask yourself what you would have if clocks didn't exist.

Quite simply, you would have two great realms of light and dark alternating with one another. Primitive humankind regarded night and day as fundamentally different phenomena and did not fuse them into a single unit. Even today, very few languages have a specific word for this most significant unit of time; in English, our word *day* is used for both the daylight part of the unit and for the whole night/day period.[1] A degree of abstraction is required to see night and day as a single unit instead of as two opposing realms. Early efforts at time-counting still did not quite deal with the whole day/night unit; instead, they usually tallied repetitions of an easily recognized event within the unit, namely, sunrise or sunset. We find this way of counting throughout the work of Homer: "This is the twelfth Dawn since I came to Ilion" (XXI).[2] In fact, both the *Iliad* and the *Odyssey* give us a feel for time-counting in a world without clocks. Particularly in the *Odyssey*, the sun seems to function as a celestial alarm clock in the morning:

"As soon as Dawn with her rose-tinted hands had lit the East, Odysseus' son put on his clothes and got up from his bed." (II) and as an ordinary sort of clock throughout the rest of the day:

"When the sun sank and night fell, my men settled down for sleep in the darkened hall." (X)

"The evening sun was shining on the fields of Ithaca when they reached the island." (XIV)[3]

The sun *was* a clock, and all the clock that was needed by Homer's warriors or by the farmers and early city-dwellers of his world. There is a reason why the calendar, with its year, month, and week, was developed so long before the clock, with its much shorter units of time. Particularly once he began to cultivate crops rather than simply hunt or gather, early man *needed* to be able to chart the seasons to know when to plant his crops and move his herds; he didn't *need* to be able to count hours within the day.

Nevertheless, the sun was always there, always casting shadows, and it was surely used for millennia as an informal index of the time of day before a device was invented to measure it. At least in sunny climates, this goldmine for time-counting lay all around—at the foot of every upright person, plant, animal, and building—a gift of the geometry of our solar system that man did not yet have any use for.

For centuries, only the sunlit hours were measured; the night hours were useless for any activity and thus there was no reason to try to count them. Looking out from our electrically lit night world, it is almost impossible for us to grasp how utterly dark night is without artificial lighting. With only the full moon and fires, and later, candles to put a tiny dint in the blackness, the night was not just a fallow realm but a fearsome one, useful only for sleep.

So let's begin with the truncated day of sunlight, and look at the first efforts to divide the time within this period. In things to do with the sun, all evidence points to Egypt, which seems always to have broken through into consciousness first. Not only did it give us the solar year, but the earliest known devices for measuring the sun's shadow have been

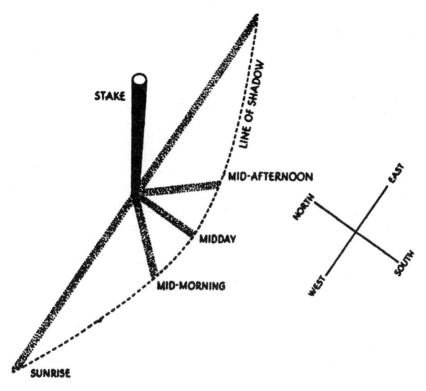

2. The first sundial was simply a stick in the ground that indicated time both by the length and direction of the sun's shadow.

found there. The very first sundial was simply an upright stake placed in the ground (called a gnomon, "to know" in Greek) so that the sun's shadow could be observed. Such a stake shows that the sun's shadow changes in both direction and length throughout the day. The farther overhead the sun is, the shorter the shadow it casts. Everyone has noticed the lengthening shadows of the late afternoon and some of us get up early enough to see the same phenomenon in the morning. And shadows of course fall to the west in the morning when the sun is in the east, and move around to the east during the afternoon when the sun has crossed into the western sky.

We don't know the exact extent of need for greater precision in time counting in the second millennium B.C., but we do know that cities and trade had become well established. The more complex life styles they bred would certainly have been made easier by a firmer awareness of the time of day. Whether it was a spontaneous invention or a long sought solution to practical need, someone found a way to improve upon the primitive gnomon by calibrating the changes in the shadows it made. A piece of such a sundial was found in Egypt and has been dated to about 1500 B.C., in the reign of Thutmose III. In the morning the upright T shape would be placed toward the east, so that its shadow would fall upon the horizontal bar. At noon, the instrument would be turned around so that the T shape was at its westward end and would continue to cast a shadow on the calibrated bar until the sun went down. By making the calibrations farther apart for the times furthest away from noon, this first "clock" dealt with the problem of lengthening shadows and was able to divide up the daylight time into approximately twelve equal parts.

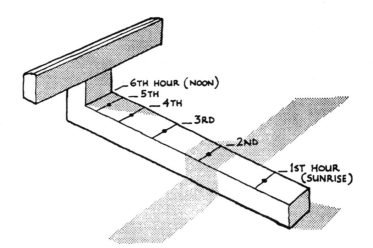

3. Ancient Egyptian sundial, c. 1500 B.C. By making the calibrations farther apart for the times furthest from noon, the dial compensated for the fact that movement of shadows slows as the sun gets higher in the sky. The direction of the instrument is reversed at noon, so that it can then count the six hours from noon until sunset.

This is the most crucial distinction between the ancient solar clock and our way of keeping time. Twelve equal *parts* of the daylight is something completely different from twelve of our equal *hours* of sixty minutes each. Our hours are the same the year round, while the "equal" in the ancient sun clock's hours meant only that all sunlit hours on a given day were the same, but each summer hour was of course much longer than a winter hour. In fact, except on the equator, every single day of the year has a different amount of daylight, so that these sundial hours are referred to as temporary hours. Think of 5:00 P.M. on a work day: If you live around the latitude of New York City, in the winter it is dark outside, while during the summer there is bright light, but it is still 5:00 P.M. in both cases. If you had lived *any where, any time* before about seven hundred years ago, this would not have been so. If the *sunlit* hours are divided into twelve equal *parts*, then 5:00 P.M. in the winter would already belong to the realm of night and would not have been counted at all before the invention of devices which use something other than the sun to count time. The twelfth hour of sunlight would have ended around 4:30 P.M. During the summer, 5:00 P.M. would only have been about the ninth of the twelve long sunlit hours. It is at least as hard for us to try to imagine this as it is for us to try, as in the previous chapter, to imagine our earth with a shorter total amount of day and night. And yet, since some form of clock time began, at least as early as the second millennium B.C., *we* are the great exception, *not* the rule.

Dating back to 1300 B.C., evidence of the first sundials that more closely resemble those with which we are familiar has been found in Egypt.[4] Like our own sundials, these instruments measured time solely by the *direction* of the sun's shadow rather than by its length. Particularly during the five centuries preceding the Christian era, many varieties were invented and many innovations made by the Greeks and Chaldeans as well as by the Egyptians. To deal with the fact that the length of the sunlit day and the sun's path across the sky (and hence the shadows it casts) vary throughout the year, dials were designed so that they could give the correct time (i.e., the correct twelfth part of the sunlit day) in any particular season, or even in any

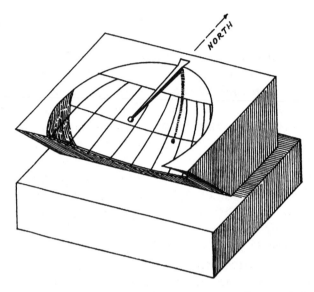

4. During the time of Alexander the Great, many forms of the concave hemispherical sundial were invented. The twelve divisions running north–south represent the temporary hours, and the lines which cross them correspond to the seasons (and in some cases even to the months).

particular month. In addition to the flat dial with which we are most familiar, sundials were also invented in the form of sunken hemispheres and cones, cubes, columns, open rings, and opened tablets, to name only a few. Already in the Roman world of the first century B.C., Vitruvius reported the existence of thirteen different kinds of sundials, and there was variety in size as well as in style. In 10 B.C. Augustus Caesar erected a giant public sundial in the Campus Martius: a huge obelisk served as the gnomon and the surrounding pavement, with inlaid lines of bronze to mark the hours, served as the dial. At the same time, wealthy Romans carried around "watches" in the form of tiny pocket sundials only a little over an inch in diameter.[5]

Nevertheless, the Romans were slow in seeing the need to break time up into hours. The Greeks had sundials a good two centuries before they were introduced into the Roman world. Indeed, sundials were a fixture of life in Athens as well as in other Greek cities from the fifth century B.C. onward, and the Greek astronomers, unlike the Romans, understood the need to tailor-make each sundial to the latitude of the city where it would be used. (The shadow cast by the gnomon is different for each latitude because the apparent path of the sun is different.) The Roman Scholar Pliny the Elder tells us the first sundial was not introduced into Rome until 264 B.C. It was a beautiful sundial from Catania in Sicily brought back as booty from the First Punic War. This sundial gave Romans the incorrect time for a century, until, in 164 B.C., the censor Q. Marcius Philippus gave Rome its first sundial correct for the Roman latitude.[6] Unlike the universal standard of time we are used to, the ancient sundials gave strictly local time.

Despite the scientific study that clearly was made by the inventors of the various types of sundials, and despite the fact that public sundials existed in Rome, the Romans' everyday world of time could not have been more different from our own. Were a Roman of the first century B.C. to time-travel for a day into our world controlled by the ubiquitous, overbearing precision of individual wrist watches tenaciously bound to public time schedules, he would think he had landed on Mars. Despite the size and complexity of the city, life in Rome "remained rural in style and in pace."[7] There is no evidence that ownership of sundials extended beyond the wealthy or that attention to its reading by the rest of the populace amounted to anything more than a quick glance if and when one happened to be near a public sundial (and if the day happened to be sunny). Sundials were *not* the watches of the ancient world.

The Roman day began just as it did in the countryside: at dawn. Thus, it was the sun itself that awakened the Romans, just as it had Homer's warriors nearly a millennium before. Typical of the many Roman complaints about this fact is Martial's whine:

"The laughter of the passing throng wakes me and Rome is at my bed's head. . . . Schoolmasters in the morning do not let you live; before daybreak, bakers. . . ."[8]

Before the sundial finally brought the concept of *hours* to Rome in the third century B.C., the sun alone was also quite adequate for marking the single punctuation of the Roman day: the moment the sun crossed the meridian at midday. A cryer in the Forum would call out when the sun reached its highest point in the sky; this signal of the turning of the day was particularly important for lawyers who had to appear at the tribunal before noon in order to validate their pleadings.[9] The arrival of the sundial certainly did not cause people to leap from this extremely casual time awareness to anything approaching our own.

Nevertheless, sundials never ceased to count off their temporary hours for the next one and a half millennia. Their way of expressing time prevailed throughout the late Roman period, the early Christian era and up until the fourteenth century. During these long centuries nearly all significant improvements in sundial design were made by the Arabs, who had first learned of the sundial from the second century B.C. Alexandrian school of the Greeks and whose innovations were then passed back to the European culture of the Middle Ages. It would only be in the Renaissance—and to satisfy a need of a machine the ancient world could not have dreamed of—that the sundial would undergo a final rebirth of its own.

Chapter III

The Measurement of the Night Hours

And the best of all ways
To lengthen our days
Is to steal a few hours from the night, my dear.
Thomas Moore

The virtues of the sundial are many, but they can never overcome its single great defect: it requires the sun. To be worthless on cloudy days (and in regions of the world that are usually cloudy), at night, and *any time* indoors is to be worthless too much of the time. As cities grew, nature loosened its grip on people's lives—they began to need a sense of the time when they were indoors, when it was cloudy, and perhaps even during the night. In order to do that they needed to find something to measure that was more reliable and more controllable than the sunlight. Over the centuries and around the world many devices were invented, all with merits and drawbacks. Sandglasses and water clocks worked by controlling the flow of sand or water. Water clocks had a problem in cold climates because the water would freeze, and even when it didn't freeze it became more viscous and flowed at a different rate when the temperature became too cold. Water could also evaporate very quickly in warmer climates.

Sandglasses weren't troubled by temperature changes, but they had other problems: sand is much coarser than water and thus can't measure as precisely. Unless the hole through which the sand flows is made of very hard material, the friction of the sand will slowly wear it down, making the hole larger and thus the flow rate faster, causing the sandglass to become inaccurate. It was also impossible to make sandglasses large enough to go through the night, so that turning them exactly when the last grain ran out presented a problem. What alarm was going to get someone up to turn the sandglass, and if such an alarm clock existed, what need would there be for the sandglass? As is the case with our use of them today, sandglasses worked best for measuring short periods of time: cooking times in the kitchen, a lawyer's speaking time in court, a preacher's time in the pulpit, a teacher's lecture in school.

Instead of using the *flow* of something to count off time, another category of device harnessed the change wrought by fire. The Saxon king Alfred the Great (849–899) is credited with inventing the use of burning candles to mark the time. Once the candle's rate of burning was determined, candles of uniform thickness could be marked with horizontal lines so that the hours could be counted as the burning wick dropped below each mark.

Both the sandglass and the burning candle are featured as metaphors for the passage of time and the brevity of life in many works belonging to the great seventeenth century genre of *vanitas* paintings. In these visual meditations on the transcience of human life, the candle's very nature—to burn down to nothing or to be snuffed out prematurely—is an obvious symbol of human life, but its meaning was surely enhanced by the fact that candles had been used for centuries as clocks.

Although Western tradition credits Alfred the Great for first using candles as timepieces, they were probably in use before then in the East, as were both sundials and water clocks. Certainly by the Sung dynasty (960–1279) both graduated candles and sticks of incense were used in China to measure time. Incense burns at a very even rate and is inexpensive, so that it has continued to be used down to the present century for measuring time. Because it comes in different aromas, different kinds of incense were used for different hours—for one hour frankincense might be used, for another, myrrh—creating a clock where the hours could literally be *smelled*.[1]

All of the fire devices had the same problem for long-term timekeeping as did the sandglass. Who was going to light the next candle or piece of incense at precisely the moment the old one ran out? Like the sandglass, they were better for timing shorter periods during waking hours.

Of all the devices invented to count time when and where the sun wasn't available, *by far* the most successful and widely used was the water clock, or clepsydra ("water thief" in Greek). Fortunately, in the ancient world the most advanced cultures were at latitudes where the temperature usually didn't drop below freezing. Once again, it is in Egypt where we find the first evidence of a water clock, and within a couple of centuries after the invention of the sundial. Water clocks were widely used by the Greeks from at least the fifth century B.C. and were introduced into the Roman world during the second century B.C. Although the Arabs made extraordinary elaborations of water clock design throughout the Middle Ages, they were first introduced to the water clock through the writings of Alexandrian astronomers and engineers. Like the sundial, the water clock existed continuously in the West from the time of the Roman Empire on through the Middles Ages. We learn from Petronius and elsewhere that it had become a status symbol in first century Rome,[2] and we also know that its water helped put out a fire in the abbey of Bury St. Edmund in the twelfth century.[3] In fact, it was probably the most accurate clock up until the pendulum clock was perfected in about 1700.[4]

Like the sandglass, the water clock was used as a simple timer to limit such things as lawyers' speeches during judicial proceedings and the guard duty of military men, but it also constituted the first successful method for counting time throughout the day and night.

The earliest form of water clock, called the outflow type, could not have been simpler: a hole was placed near the bottom of a large bowl upon whose inside wall (or outside wall if it was glass) horizontal lines were drawn at hour intervals. As the water dripped out through the hole, the water level dropped from hour line to hour line. The more precise inflow type of water clock consisted of two basins: water was added continuously to the first basin (to keep the water level the same and thus its downward pressure constant), which also had a hole near its bottom out of which water flowed at a constant rate, but this time it passed into a second vessel marked with gradations for the hours. As the water rose in the second vessel, it marked off the passing time.

Often the inflow type of clock was rigged with devices which acted as alarms by dumping metal balls or pebbles, or causing whistles to blow when the water reached the top of the inflow vessel. Tradition has it that Plato (427–347 B.C.) invented one of these "alarm clocks" to awaken his students at

5. Plato's inflow type water alarm clock. Water entered the (usually) graded vessel (b) at a constant rate. The vessel filled during the night. When the water reached the top, it tipped over a bowl containing lead balls (a), which was hinged to the top of the vessel. The balls then fell onto the copper platter (c) below, and woke up Plato's students.

the Academy he founded in 378 B.C. He placed a huge inflow type water clock in an olive grove beside his pupils' dormitory. A hinged bowl containing metal balls was attached to the top of the inflow vessel, and when the water reached it, its force caused the bowl to lift and tip over, dumping the balls onto a copper plate below. This made enough racket to awaken at least some of the students, who could then wake up the others.[5]

In speaking of Plato's device, we are talking about a clock that went on through the night, as do our own. With the water clock, we are thus brought a tiny step closer to our own way of measuring all of the time within one cycle of the earth's rotation, but we are still a world away, both conceptually and technologically. Although they are now tied by a slender thread, the sunlit and dark hours remain two separate, opposed realms, just as they were for Homer. In order to see why this was so, we need to look at the relationship between sundials and water clocks.

In the Roman agora in Athens can still be seen today the remains of the Tower of the Winds, a magnificent first century B.C. sundial. Or, rather, sundials—there were eight of them, each facing in the direction of one of the eight winds identified by the Greeks. However, what interests us now are not the sundials but the cistern beside the Tower, which served as a reservoir for the accompanying water clock. Similarly, in the second century B.C., the censors who succeeded Q. Marcius Philippus, not wanting to be outdone by him, contributed to Rome a water clock that was placed beside the sundial Marcius had donated. These water clocks placed beside the sundials did service on cloudy days as well as at night, when a reading couldn't be gotten from the sundial.

This physical bonding of public water clocks to sundials is a clue to their reciprocal relationship. But forget for a moment these sundial/water clock pairs and consider the technology of the two instruments. At least in antiquity, it was much simpler to make a sundial that broke time up into equal parts of sunlight rather than into our own equal hours, but the same cannot be said for water clocks. In fact, the water clock would seem to be an ideal instrument for measuring equal hours: the vessel could simply be calibrated with equidistant lines to measure off equal periods of time. Indeed, some water clocks did measure equal hours, but the vast majority did not, either in the ancient world or in the Middle Ages. We imagine that it would have been so easy to start right off with our twenty-four equal hours, rather than wait another two millennia for them, and the water clock was the instrument with which it could have been done.

That is not what happened. People in the ancient world were not ready to abstract time from the central reality of the sun's daily passage, and so the aspect of innovation in using water clocks to tell the time throughout the night was countered by the fact that there was no innovation at all in their manner of doing it. The calibrations marked on the water clocks set beside sundials are not the simple equidistant lines that would make sense to us; rather, they are a translation into water clock format of the readings given by the sundial. The procedure was really quite straightforward. First, the sundial's readings were copied onto the water clock so that it could be used on cloudy days: Whenever the shadow cast by the gnomon arrived at one of the sundial's hour demarcations, a mark would be made on the side of the water clock at the level of the water at that moment. This would be continued throughout the day until the water clock had twelve marks indicating the water levels corresponding to each of the sundial's temporary hours.

Since these temporary hours change throughout the seasons, a way had also been found to modify sundials so that they could show the correct time for the season and even for the month, and a way was found to do this on water clocks as well. What of the night hours? The data were simply reversed, because the hours of darkness are always the antithesis of the daylight hours. The marks copied from the sundial for, say, July 1, are equally accurate for the night hours of January 1. This is the case for any two days that are six months apart. (Only at the vernal and autumnal equinoxes are the day and night hours the same.)

Thus, although the water clock was the first instrument invented which was capable of measuring time throughout the day and night, it did so by preserving the primordial dichotomy between the night and day. Although Arabic engineers made elaborate refinements of the water clock which were subsequently transmitted to medieval Europe, they never touched this dualism of night and day. Slate fragments found at the Cistercian Abbey of Villers in Brabant and dated to 1267 or 1268 are inscribed with instructions on how to set a water clock for *every single day of the year* according to the varying time periods of light and darkness![6]

In focusing on the water clock's technology, it is easy to lose sight of the fact that its actual role in people's daily life was extremely peripheral. If the sundial was assuredly not the watch of earlier cultures, the water clock was even less so. In the first place, it wasn't portable—sloshing water doesn't line up well with hour marks. Second, water clocks never got beyond being expensive status symbols for the ancient world's wealthy, and during the Middle Ages they were mostly confined to monasteries. In this regard, it is helpful to recall the words of Marc Bloch concerning the time awareness of people in the Middle Ages: ". . . the passage of time escaped their grasp all the more because they were so ill-equipped to measure it. Water clocks, which were costly and cumbersome, were very rare. Hour-glasses were little used. The inadequacy of sundials, especially under skies quickly clouded over, was notorious."[7] There is one last chapter in the reign of the temporary hour yet to come, but remember the dates of the Villers Abbey fragments (1267–1268) because they are close, so very close, to the great divide that rises between us and the ancient and medieval worlds regarding time awareness. With our clocks ticking away their invariant sixty-minute hours, we can try to imagine, but we can never reenter that lost world when the hours stretched and shrunk to keep time with the journey of the sun across the sky.

Chapter IV

The Canonical Hours

Idleness is the enemy of the soul.
St. Benedict (480–543)

Ideas about time are governed at least as much by cultural beliefs as they are by scientific knowledge. The dominant feature of medieval culture was the Christian Church, to the point that there was almost no such thing as a purely secular activity. Every occupation had its patron saint, every castle its attached chapel, and the calendar year was a forest of saint's days and religious festivals. The Church also developed a way of ordering the day's time that enveloped people's lives just as profoundly as did the ecclesiastical activities. This new system of time was created within the context of the temporary hours, but its allegiance was to something other than the physical cycle of night and day.

Like Christianity, the other great monotheistic religions of the West (Judaism and Islam) also had times of the day that were set aside for worship. Judaism called for prayer three times during the day and Islam for five. But in both cases, the times for prayer were broad, and they were closely associated with major events in the sun's cycle. The pious Jew of the Middle Ages was required to pray during the morning, afternoon, and evening, defined as after daybreak, before sunset, and after dark. Similarly, in Islam the obligatory prayer times were at dawn, right after the sun reached its meridian at midday, before and just after sunset, and after the fall of darkness. These time periods gave the believer a lot of leeway for a personal act that in neither religion required the intervention of clergy or fellow congregants. ". . . In Islam as in Judaism the times of prayer are bands rather than points."[1]

The situation with Christianity was different. Already in the late second century, the North African priest Tertullian (160?–?230) added to the Jewish prayer times (upon which the early Christians built and from which they subsequently deviated) the requirement that Christians pray at the third, sixth, and ninth hours of the day. The Romans had been slow in adapting clock time to their daily lives, but by the third century the habit of dividing the day into quarters was well established, and the third, sixth and ninth hours were announced publicly. Thus, the times for prayer of the early Christians were already set by the secular time points of the Roman business world and thus could be easily followed by the Christian faithful.[2] The point here is that from early on the Christian habit of worship was geared to the clock rather than to the sun's cycle. However, in saying this we must remember that throughout the long continuum from sun time to clock time there are many subtle distinctions. The Christian clock was of course the ancient clock of elastic temporary hours, and in that sense it was still deeply embedded in the natural cycle of the sun. Nevertheless, there is a categorical difference between praying, say, after sundown and praying at the ninth hour of the day—the latter is an invented time point while the former is still tied to a specific event in the natural cycle.

During the early centuries of Christianity the hours of worship and the prayers recited varied greatly from group to group according to local custom and habit. Although there would always be variations, a standard was set in the sixth century by St. Benedict (480?-?543), who codified seven precise times for prayer as part of the rules for the liturgical day of his monastic order: lauds (just before daybreak), prime (just after daybreak), terce (third hour), sext (sixth hour), nones (ninth hour), vespers (eleventh hour), and compline (after sunset); an additional time, matins, was for prayers said during the night.[3] The Psalms and scripture readings to be said at these times were known as the Divine Office, and the prayer times became designated as the Canonical Hours. During the following centuries this schedule for the reciting of the Divine Office spread across Europe along with the powerful Benedictine order, and it was later adopted by other monastic orders.

The day of a medieval monk, punctuated by so many calls to prayer, would seem at first glance to be at least as time-conscious as our own. Looking more closely, we notice that the canonical hours are not evenly spaced, so that in terms of rationally counting time itself, the canonical hours actually represent a backward step from the straightforward neutral temporary hours of the ancient world. In actuality, the canonical hours were alarms going off throughout the day, and the monks' awareness was more of the alarms for worship than of time counting per se. And how did they know when a particular hour had arrived? They certainly couldn't set the alarms on their wrist watches.

They knew by bells. Bells, bells, bells, and more bells. Bells have been around forever—China was making bells in its Bronze Age (before 3000 B.C.) and bell-making in one form or another has continued throughout the world up until the present day. But never—before or since—have bells played a more central role in human life than they played in medieval Europe. The use of bells in connection with Christian ritual began as early as the sixth century, and the ringing of bells to mark the canonical hours is believed to have been instituted by Pope Sabinius in the early seventh century. Benedictine bell foundries existed by the eighth century, and by the late tenth century, bells had become a part of the liturgical service itself. Long after the advent of the mechanical clock in the late thirteenth century, bells continued to be the public time-pieces of the medieval world. Our word *clock* derives from the word for bell in medieval Latin (*clocca*), as well as in all of the European languages: *clok* in Middle English, *clocke* in Middle German, and *cloque* in Old French.

If the bells told the people what to do, what told the bells when to ring? What else was there but sundials, water clocks, and stars? At least during the early Middle Ages, sundials were far more primitive than their ancient counterparts. The obviously named scratch dials, which in England date as far back as A.D. 670, were literally scratched into the stone on the south walls of churches. These primitive instruments were marked in a variety of different segments of only approximately equal length. Some have been found with lines unequally spaced, and it has been conjectured that they were used for timing the canonical hours.[4]

The need for water clocks, particularly those equipped with alarms, can easily be realized when we look at the only other available method for determining the times of matins and compline, which occurred at night. These were the instructions for the bell-ringing monk at a monastery near Orleans during the eleventh century:

> On Christmas Day, when you see the Twins lying, as it were, on the dormitory, and Orion over the chapel of All Saints, prepare to ring the bell. And on Jan. 1st, when the bright star in the knee of Artophilax (i.e. arcturus in Bootes) is level with the space between the first and second window of the dormitory and lying as it were on the summit of the roof, then go and light the lamps.[5]

Let's take a moment here to reflect on the curious course of human knowledge. In the twentieth century, religion is not generally thought of as an eager champion of progress in scientific knowledge, and many people, recalling such events as Galileo's trial for heresy, believe that historically religion has been somewhat of a reactionary force where secular knowledge is concerned. But in the Middle Ages, before the rise of universities in the twelfth century, the monasteries were without question the central places in society where efforts were made to keep the ancient world's knowledge from oblivion and where new knowledge had a chance of being acquired. The rigid adherence to repetitive prayers said many times a day may seem to be far from a stimulating exercise, but, as the above instructions tell us, as a byproduct it kept alive such ancient learning as knowledge of the stars—their names, their paths, their times of rising and setting.[6] It was also the impetus without which many ingenious forms of water clock alarms might never have been invented. The previously mentioned notations at Villers Abbey giving the hours for every single day of the year were for the express purpose of accurate alarm ringing for prayers, but in the process they kept alive habits of numerical exactitude and knowledge about the sun's path.

In terms of sophistication, the medieval life of the water clock was the reverse of that of the sundial. Whether the medieval water clock represented an uninterrupted continuation of the ancient Roman technology, or whether that tradition was lost during the chaotic centuries after the fall of Rome and only reemerged through translations of Islamic treatises, is not known. What *is* known is that medieval water clocks were complex instruments whose primary function was not to measure time so much as to sound alarms so that the bell ringers could sleep rather than watch the sky. The regulations of the Cistercian order contain several rules which apply specifically to the care and daily adjustment of the water clock. Although their mechanisms were more sophisticated, in principle these monastic clocks were analogous to Plato's ancient alarm clock which went through the night and sounded an alarm at a precise moment in the morning. But instead of Plato's pebbles falling into a metal plate, the alarm that went off on the medieval clock was—you guessed it—*little* bells.

It was not only the sound of the bells throughout the town and countryside that spread consciousness of the canonical hours far beyond the monastery walls. The ordinary people of the Middle Ages had an intense religious piety, and their complaints about the dissoluteness of some of the Benedictine monasteries played a great role in efforts at reform as well as in the establishment of the new, more ascetic orders. The Carthusian and Cistercian orders were established in the late eleventh century, and by the twelfth century even more monastic orders had emerged, all of them characterized by an ascetic, individual mode of worship in deep contrast to the self-sufficient, wealthy, communal organization of the Benedictine order. This shift from the community to the individual culminated in the Dominican and Franciscan orders, founded in the early thirteenth century. Friars from these orders took vows of poverty and often traveled and lived alone. So that they could recite the Divine Office on their own, books called breviaries were created containing the Psalms, prayers, and scripture readings to be said at each canonical hour.

The piety of the laity also gained new value with the movement away from strictly adhered-to communal rites and toward the personal experience of private prayer. Now that religious observance could be a matter of private experience, lay people wanted something for themselves that resembled the breviary of the itinerant friars. Thus was created the Book of Hours, the lay breviary, containing readings devoted to the Virgin Mary that the owner could read or recite at the appropriate canonical hour. Its use in England has been documented in the mid thirteenth century, but it was in the fifteenth and early sixteenth centuries that it reached its greatest popularity on the Continent. The Book of Hours was the best-selling text of the later Middle Ages, and the great demand for these books by the wealthy was a crucial force behind the development of the Gothic illuminated manuscript.

Quite aside from what it says about the medieval world's piety, the Book of Hours was certainly a vehicle for increasing—and also internalizing—the sense of time discipline among the laity. It may also serve as a compendium of medieval ideas about time. If we sit back and look at any Book of Hours long enough, we slowly become aware that it is a paradigm of the ways time was structured in the medieval world. The backbone of the Book is of course the canonical hours, the godly clock that punctuated the day of the faithful. These hours, irrational as divisions of sunlight, were not geared to the natural world but to God. And yet, as we look at almost *any* illuminated page of almost *any* Book of Hours, we suddenly realize that these books teem with the most loving images of the natural world. Borders are decorated with leaves, flowers, birds, and butterflies. The bright colors of our world dazzle in every scene and, in the later books, exquisite lilliputian landscapes surround the figures. Most important, by the fifteenth century most Books of Hours opened with images of the Labors of the Months in connection with the calendar of church feasts and festivals honoring saints. What are these images of nature and rural tasks throughout the year but a depiction of calendar time? Calendar time long preceded clock time because agrarian activities did not require any clock other than the sun, and here we see its powerful continuance. By the early twelfth century there was considerable urban life in northern Italy and the Low Countries, but throughout the Middle Ages the vast majority of people still lived on the land. We may talk of sundials and water clocks,

6. Page from *The Tres Belles Heures of Notre Dame*. This Book of Hours was listed in the 1402 inventory of Jean de Berry's manuscripts.

canonical hours and ringing bells, but for the rural mass of medieval people, time still moved not by divisions within the day but to the rhythm of the great slow round of the seasons. The bells, loud as they were, were just a tinkle against the booming pageant of nature's cycle of death and regeneration.

That was to change very soon, and both of these medieval structures of time were to begin to fade in the bright light of a new time. Within the Book of Hours they still reign. "The various times within Time"[7] will always run along beside each other, overlap, and never be resolved into any grand unity. (The most glorious Books of Hours were produced during the fifteenth century, long after the advent of the mechanical clock with its new time of equal hours.)

Yet another idea about time lies enfolded within the pages of the Book of Hours: the belief that God created the world and everything in it—including time. As we will see in the next chapter, the ramifications of this theological idea in the economic world and the battle which ensued between the forces moving the world toward more rational, secular activity and those loyal to the theological idea that time belongs to God reveals something of the watershed in time perception that was fast approaching.

Chapter V

The Selling of Time

Time is a gift of God and therefore cannot be sold.
Medieval saying

Time is money.
Benjamin Franklin

Before we cross the great divide into the early years of our own way of perceiving—and counting—the hours of the day, we need to make a slight detour to look at another ramification of the medieval belief that time belongs to God: Since time is God's (or God's gift to humankind), we must not buy or sell it. What this detour lacks in information about the means of telling time, it will hopefully make up for by clarifying the attitudes which underlie those means.

Nothing in our lives seems further removed from theology than our finances—we may not enjoy dealing with their harsh realities, but nothing seems more this-worldly and quotidian-bound. Mortgage payments, car payments, student loans, retirement plans, and decisions about where to invest whatever money is left over surely have absolutely *nothing* to do with religion. And yet what are all of these transactions but the buying and selling of *time*? A mortgage costs a lot more than paying up front because we're buying the house *plus* the time we get to live in it before we've paid for it. Conversely, when we buy a treasury bill we're loaning the government the use of our money for a period of time; the bill is worth more when it comes due because the government has paid us interest for the time it used our money. In fact, the entire apparatus of our world's economy would instantly grind to a halt if time were disallowed as a commodity to be bought and sold. The ability to buy on credit is just about as crucial for our way of life as are the hours our clocks give us.

Just as our equal hours did not yet exist in the Middle Ages, neither did the rational, secular use of time as a sellable commodity (without which there could be no capitalism). If we use the word *usury* today at all, we mean an exorbitant rate of interest, but in the Middle Ages it meant receiving *any* return whatsoever beyond the thing loaned. Thus, our whole economy is rooted in and utterly dependent upon something that was rigorously forbidden by the medieval Christian church. The story of when and how and why the change took place is extremely complicated and contains many facets which are not related to time. We can do no more here than to sketch a brief outline of the time-related aspects of this crucial transformation.

Although there were other arguments against usury as well, the argument concerning time was central. In the literature of the Middle Ages many statements can be found that express the Church's reasoning, and as good as any might be that of the author of the *Tabula exemplorum*, written in the late thirteenth century:

> Since usurers sell nothing other than the hope of money, that is, time, they are selling the day and the night. But the day is the time of light and the night is the time of rest; therefore they are selling eternal light and rest.[1]

Similar arguments were also made against payment to teachers: knowledge also belongs to God, so how can it be sold? Before deciding that such thinking is utterly foreign to us, recall our own slogan "The best things in life are free." Although we certainly don't all conceive of the natural world as belonging to God, most of us nevertheless have clear ideas about what can and cannot be bought and sold. If we were asked to pay for the air above our heads, our reasoning against it might not be so different from that of our forebears when payment for time was in question.

This medieval view had not always been held, and in fact the charging of interest on loans has a long history. Both the Greeks and Romans allowed it, although at various times both made laws setting maximum rates and Roman law specifically forbade compound interest. When Constantine the Great (272–337) converted to Christianity in A.D. 313, he made no change in the Roman laws concerning interest, and Justinian (527–565) merely lowered the maximum rates from twelve percent to six percent for business loans and four percent for nonbusiness loans.

Ironically, the Christian proscription on interest originated in the Torah, which forbade Jews from charging interest to fellow Jews (but not to foreigners). It was at the Council of Nicaea (A.D. 325) that the Christian church decided that clerics could not charge interest. This was followed in A.D. 444 by Pope Leo I's papal ban on the taking of interest by *all* Christians, whether churchmen or laity. There wasn't much change in the usury laws—nor was there need for it—from that time until the late eleventh century. Agriculture had become the basis of economic life, most manors were self-sufficient but had little surplus for trade, and the need for credit was relatively low.

This shift toward an economy dominated by agriculture occurred after the fall of the Roman Empire. Throughout the fifth century, Italy was wracked by successive invasions of Germanic tribes, but the death blow to the more diverse economy built up during the Empire came in the sixth century, when Justinian's forces battled for thirty years to win back Rome from the Ostrogoths. The fact that they were finally successful did nothing to prevent the collapse of the Roman economy. The great cities—Rome, Milan, Naples—were depleted of population as people moved back into the countryside. Both in people's minds and in reality, land and agriculture became the basis of the economy and the only true wealth.[2] It would not be until the tenth century that Italy would even begin to regain its traditional role of leader in Europe's culture and economy.[3]

Religious support for this natural economy was found in the belief that man's work ought to result in the creation of something material, in imitation of God's work of creating the world.[4] This literally materialistic attitude helps to explain the complete acquiescence of most people in the Church's sanction against usury, which produced nothing tangible. There were also many safeguards in the medieval economy to prevent competition and therefore profit, so that usury, with its goal of monetary profit, was viewed with disapproval.[5] In other words, the agrarian economy recoiled defensively in the face of an act by city people that, using sterile metal, sought profit without labor through the selling of God-given time!

Dante immortalized this attitude in the seventeenth Canto of his *Inferno*, completed shortly before his death in 1321. He placed usurers in the seventh circle of Hell:

> Their eyes burst with their grief; their smoking hands
> jerked about their bodies, warding off
> now the flames and now the burning sands.
>
> Dogs in summer bit by fleas and gadflies,
> jerking their snouts about, twitching their paws
> now here, now there, behave no otherwise.
>
> I examined several faces there among
> that sooty throng, and I saw none I knew;
> but I observed that from each neck there hung
>
> an enormous purse, each marked with its own beast
> and its own colors like a coat of arms.
> On these their streaming eyes appeared to feast.[6]

Before we ask how this mind set was slowly undermined and the way opened up for the beginnings of our own view of money and time, we need to realize that the selling of time was not problematic for Christians alone. Just as it retains a pure lunar calendar of 354 or 355 days which circles endlessly through the solar year, so also the religion of Islam has never departed from its own ban on usury. The purchase of a western-style mortgage on a house cannot be made by an orthodox Muslim even today.

The impact of the concept of usury on the lives of Jews was, sadly, of a different nature. Jews had traditionally been city-dwelling merchants, artisans, and bankers rather than farmers, and as such they were very useful to the agrarian society of early medieval Europe. Their religious law allowed them to lend money to non-Jews, and their services were badly needed by governments as well as by international trade. (The Christians somehow found less fault with being on the paying rather than the receiving end of a usurious transaction.) Trouble began with the surge of popular Christian piety in the eleventh and twelfth centuries, which spawned anti-Semitism along with the Book of Hours. Not only were Jews not allowed to hold land, but the new guilds of Christian merchants also excluded their Jewish competitors. By the twelfth century, usury was one of the few ways a Jew could earn a living.[7] The Christian antagonism to the selling of time thus had everything to do with the rise of medieval anti-Semitism, which reached its apex in the pogroms against Jews by Christians on their way to do God's work in the Crusades. It is ironic and sad that Christians got their original idea for this proscription on the sale of time from the Jewish Torah.

How did the medieval attitude toward time, and with it the stringent usury laws, finally begin to change? The two went hand in hand, their mutual fates sealed not by theology or philosophy, but by commerce and industry. From the tenth to the twelfth century, western Europe went from being far less developed than its Byzantine and Islamic neighbors to an economic and technological level that surpassed them both.[8] The European population burgeoned, many people left the countryside to return to their neglected cities or to build new ones, and a new class of people—the bourgeois—grew up; these people lived in towns and made their wealth from trade rather than agriculture.

Trade, which had been quite local within Europe and, with the exceptions of the Swedes and the Venetians, almost nonexistent with the Byzantine and Arab worlds, developed rapidly. By the end of the eleventh century Genoa and Pisa had set up colonies in North Africa for the purpose of trading with the Muslims. Within Europe, every town had a market and the huge international trade fairs on France's plain of Champagne during the twelfth and thirteenth centuries were *real* trade fairs (not like our World's Fair of today, which, like our agricultural fairs, derives from them).

Throughout this economic transformation, the Church continued to inveigh against usury. As the economy improved, there was more need for loans and thus a more powerful incentive for finding ways around the usury laws, which led in its turn to stricter, more detailed laws and tougher penalties for breaking them. The Third Lateran Council (1179) decreed that all usurers would be excommunicated. The Church couldn't accept gifts from usurers, and a conviction for usury precluded the holding of public office or testifying in court; the wills of usurers were invalid, and such people could not be buried on sacred ground. The clergy continued to preach against usury as late as the seventeenth century. Theologians at the University of Paris issued an edict against it in 1530, and Martin Luther also denounced the practice of usury. In capitalistic Holland as late as 1640, Calvinist ministers still preached against the sin of taking interest for the loan of money.[9] Nevertheless, beneath the ecclesiastical harangues, subtle changes were taking place as early as the twelfth century.

Medieval lawyers and their clients became spectacularly adept at circumventing the laws by disguising interest payments. The church itself was a borrower (and occasionally a lender), and it, too, made use of the ingenious methods of casuistry that had been developed for paying interest without appearing to pay interest. In short, credit financing had become too pervasive and integral a part of economic life to be successfully fought against. It was a fact of life—and an economically productive fact of life—that no amount of theological argument was going to make go away. Although theologians and scholars continued to argue the moral fine points of the usury problem, by the mid fourteenth century there was a marked decrease in the Church's actual prosecutions for usury, and it even began to change its laws to allow moderate interest rates to be legally charged.

As Jacques Le Goff has shown, one other indication that the Church had softened its stance on the usury issue is found in the Confessors' manuals. These manuals, which were guidebooks created to help priests mete out the appropriate acts of penitence, began to proliferate after the Fourth Lateran Council of 1215, which decreed that all Christians must go to confession at least once a year. The act of confession in itself created a meeting ground between the Church and its money-dealing merchant parishioners, and forced them to come to some kind of terms. Where they and their usurious practices had been so vehemently denigrated by the Church only a short time ago, already in the thirteenth century these manuals reveal that the Church had begun to accept the fact that the buying and selling of time was the life-blood of a large segment of its flock.[10]

The same economic forces that slowly brought about a quiet death to the proscription on usury also created in their wake the beginnings of a new attitude toward time. Although it may not be immediately obvious, the kinds of commercial networks and transactions that grew up simply could not have existed if the only ways people thought about time were in terms of the daily and annual cycles of nature or of the Church's canonical hours and religious festivals. These structures of time remained the primary frameworks for the vast majority of people, but the merchants' activities introduced a third way of perceiving time: time as normal, regular, predictable, and abstracted from *events*.[11]

We are not yet talking about mechanical clocks on the plain of Champagne in the twelfth and thirteenth centuries. We are not even talking about equal hours. Before there was a clock to measure it, the activity of conducting business had already begun to create a *sense* of time that was rationalized and abstracted from the events of nature or the Church's hours of worship. As time became rationalized, it necessarily grew to be perceived as more secular. This secular time did not take the place of the other temporal frameworks; it simply existed alongside them, and people moved from one mental framework to another with ease (and without hypocrisy).[12] As the Church accommodated to the realities of its congregants' lives on the matter of usury, so God's time began to grant space to the new secularized idea of time required by a money economy.

Chapter VI

The Mechanical Clock: The Product

Each hour is a little bucket of time to fill.
The clock says when to pour.

John Boslough

For all the talk in the previous chapter about the new sense of time felt by people engaged in commerce, and for all the monastic concern for the temporal regimentation of life, the fact of the matter is that these were just two small segments of the population and that we are still in the inchoate realm of attitudes and sensings rather than in that of concrete time reckoning. The merchants and monks certainly created a desire to track time more closely than sundials and water-clocks were able to, but to make that leap required a new kind of machine.

We have all heard great claims for the mechanical clock. Lewis Mumford: "The clock, not the steam engine, is the key-machine of the modern industrial age,"[1] and " . . . the clock was the most influential of machines, mechanically as well as socially."[2] David S. Landes: "This was one of the great inventions in the history of mankind—not in a class with fire and the wheel, but comparable to movable type in its revolutionary implications for cultural values, technological change, social and political organization, and personality."[3] If we are honest, most of us will admit that we just don't quite see it. We easily see the great claims for the printing press, for electricity, for the automobile, airplane, and a host of other inventions, but we just don't quite get the supreme importance of the mechanical clock. The explanation for our dimness lies at least in part in the fact that the time produced by our clocks is so close to us, so deeply a part of us, that it's hard to even think of it as an invention. In fact, most of us think of it as time itself. We have all had moments of trying to imagine what life would be like without these other inventions, but our rootedness in the peculiar sense of time created by our clocks is so ingrained that we are probably incapable of conceiving what time would feel like in a world without clocks.

For example, our clothes made of natural fibers have gone through a complex manufacturing process, as has the wood of our furniture. Their basic "stuff" is natural, but their form is not. We have no trouble seeing this. And yet we must say of the time our clocks give us exactly what we say of our natural fiber clothing or our wooden furniture: only the "stuff" is natural, not the form. We could no more find our clock time in nature than we could find a wooden chair sprouting from the soil or a sweater growing from a sheep's back.

The clock is a machine and, like every other machine, it produces something. The specific products made by clocks are hours (twenty-four of them each day), minutes (sixty in each hour), and seconds (sixty in each minute).[4] Unlike the temporary hours produced by sundials and water clocks, the mechanical clock's time segments are always the same, winter and summer, day in and day out. There is not one thing natural about them (with the crucial exception of the whole twenty-four-hour unit, which is the amount of time required for the earth to turn upon its axis). Within, the divisions are pure conventions—they might have been anything.

The mechanical clock, at only seven hundred years old, is one of our most recent inventions for time measurement, but the pieces into which it cuts time go back to the ancient world. Although their hours were of course the expanding and contracting temporary hours, the convention of twenty-four hours in the day/night cycle nevertheless began in ancient Egypt, while the sixty divisions of our minutes and seconds derive from the number system based on sixty (sexagesimal) of the Mesopotamians. (Because they had not yet invented fractional numbers, the Mesopotamians favored whole numbers which could be divided in several different ways, and the number 60 can be evenly divided by 2, 3, 4, 5, 6, 10, 12, 15, 20, and 30.)[5] Had history gone differently, our time system might have been the decimal (based on ten) which we use for most of our other measurements, or the digital (based on two) of our computers, or even the vigesimal system (based on twenty) of the Maya, but what lives on in our clock time is the sexagesimal system of the ancient Fertile Crescent.

For a brief period after the French Revolution, from 1793 to 1795, the people of France lived according to decimal time:

ten equal hours for each day/night cycle, one hundred minutes in each hour, and one hundred seconds in each minute. The leaders of the Revolution wished to create a truly new world, and toward this end no stronger gesture can be made than to change the very beat of time. The decimal divisions of time were every bit as good as the sexagesimal; what they lacked was habit. So difficult was it for people to make the transition that clocks were made which showed both decimal time and sexagesimal time. (Nothing reveals more clearly than this surreal machine, made for the purpose of comparing one fictive grid with another, how often "reality" gets located

7. Such French Revolutionary period timepieces were made during the two years in which revolutionary time reigned (1793–1795). This watch shows the twenty-four hours (twice twelve) of our day alongside the ten hours into which the Revolutionaries divided one rotation of the earth. One of their hours was thus nearly two and a half times longer than one of ours.

in our own conceptions rather than in the elements they were meant to demarcate.) In 1795, after less than three years, the Revolutionaries gave up, and France returned to the traditional fiction of the sixty-minute hour shared by its European neighbors.

Clearly, what was new about clock time was not the numerical system upon which it was based. What, then, was it that made the mechanical clock's time so revolutionary? The answer begins and ends with the fact that the mechanical clock lifted time out of its relationship to the rhythms of nature and made it something abstract and autonomous, an entity unto itself.

It did this simply by making the pieces of time itself, so that they were dissociated from anything in nature. Sundials are just the opposite. They are really just little recording devices embedded in nature. They wait upon nature, much as a seismograph waits within a mountain to register an earthquake if, and only if, it happens. No sun, no sundial recording. To put it another way, they are merely puppets which act or lay limp at the whim of the only machine involved, the "machine" of nature. True, even though its water can freeze, the water clock is nevertheless a small step removed from the sundial's absolute dependence upon the vicissitudes of nature for its reading of time. But the water clock, like the sundial, records the temporary hours of sunlight and darkness, and thus declares a conceptual entrenchment in nature that more than counterbalances its small degree of physical independence.

So powerful was the belief that time somehow inhered in the rhythms of nature that it had kept timekeeping in bondage to our planet's cycle of light and darkness for over three millennia. There is nothing surprising about this extreme reluctance to ignore the signs of nature in the telling of time. Just as our eye's plain vision says that the sun revolves about the earth, so nearly all of nature's time counters tell us that time is an elastic, organic, fluid phenomenon that beats to their beat. The natural rhythms have a regularity, but they are also full of variation: not only does the sunlit day vary in length throughout the year, but animals have different gestation periods, and crops need to be sown and harvested at different times. In terms of our own bodies, heart-rate and breathing change literally from hour to hour, depending on physical activity and emotions.[6] Only the invariant movement of the stars offers some precedent for the steady, rigid tick–tock of the clock.

The spell of this millennia-old perception of time was broken not by a raw conceptual leap but by the mechanical clock, which was technically incompatible with the making of unequal hours. The machine simply couldn't be made to do it without great difficulty. Had it been able to, there is no telling how much longer the temporary hours might have persisted.

So a new kind of time was invented by a new machine. So what? How did this transform the world? Let us first look at its effect on everyday life. Before the clock, the central unit of time was the whole day, and people drifted through it, accomplishing their tasks to the rhythm of light and dark. The sundial or waterclock served only to highlight the signs of nature. Within decades after the mechanical clock's invention, the *hour* was already beginning to replace the day as the basic unit of labor time in the medieval world's largest industry, textile manufacture.[8] By making time abstract and then cutting it up into hour-sized pieces, the clock suddenly made of time something that could be *used*, rather than merely lived through. And out of this increased awareness of time and improved control over it, grew further inventions to enhance the hours of life. To quote Lewis Mumford:

"When one thinks of the day as an abstract span of time, one does not go to bed with the chickens on a winter's night: one invents wicks, chimneys, lamps, gaslights, electric lamps, so as to use all the hours belonging to the day. When one thinks of time, not as a sequence of experiences, but as a collection of hours, minutes, and seconds, the habits of adding time and saving time come into existence."[9]

The clock did more than just raise the time consciousness of individual people and help them to keep their own time from slip-sliding away. Beginning with the fourteenth century's communal clock in the bell tower and continuing on to our own wrist watches, this new mechanical time allowed people to synchronize their efforts and gain a far greater collective efficiency. Despite all the ills that remain in our world, no one would deny that most people's level of creature comfort is light years beyond that which existed only a couple of centuries ago. We look to inventions as explanation, but the inventions alone would never have given us the millions and billions of their replicated products. For that, assembly in factories and distribution throughout the world were required. Without the synchronization of vast numbers of people's time, these inventions could no more be ours than had they never been imagined.

The effect of clock time reached much farther than the regimentation of society's time. It also "helped create the belief in an independent world of mathematically measurable sequences: the special world of science."[10] Look at the scientific experiments verified by abstract quantitative measurements that are carried out everyday in laboratories throughout the world. All of them achieve their results by looking at their subjects through the impartial measuring rod of mathematics. The liberation of time from the events of nature and its division into equal mathematical units was an early instance of this world of science. Romantics who denigrate this world as cold and impersonal should look down at their own wrist watches and realize that they, too, partake of it; whenever we check our watches we are all scientists, however fledgling, in the laboratory of our planet.

Everything about our clock time is paradoxical. It is both freedom-giver and tyrant. It is abstract, and yet at times its intricate web seems as concretely constricting as a prison. On the one hand it is unnatural and on the other, it is *more* natural, i.e., closer to the truth of nature. On the side of its unnaturalness, there is the vast majority of humankind who lived their lives not only without clock time, but without the remotest inkling of it. It was invented in the towns of western Europe in the late thirteenth century and has since spread across the whole world, but every earlier civilization accomplished what it did without clock time. It is an oddity more than a norm, and it began as an oddity. As we have seen, the extreme concern for temporal discipline of the Christian monks—the earliest constituency for more accurate clocks—was not at all native to humankind.[11] These onclaves of celibate men rising in the middle of the night to say prayers and marching through their daylight hours to the tune of more alarms for more prayers were an anomaly relative to the vast majority of civilized life before or since. Above all, our clock time seems unnatural because it makes the tick of a minute hand on a man-made instrument seem more valid, more *real*, than any sign from nature.

On the other hand, we certainly believe that our concept of time as abstract is closer to the truth than is the idea of time as embedded in the natural rhythms, and the world of science that has developed from this abstract concept surely indicates as much. But it will happen—it already has—that our abstract concept will be superseded by yet another that more closely approaches nature's reality. According to relativity theory, time is *not* abstract—in the sense of being absolute and unrelated to anything external to it. It is *not* a consistently flowing entity unto itself, but a phenomenon that is warped and stretched by the effects of matter's gravitational pull in the field of space-time. In fact, it has been experimentally proven that at high velocities time *literally* slows down. If a clock were put on a spaceship traveling at eighty-seven percent the speed of light, it would record half as many hours as would a clock on earth. If that spaceship could reach the speed of light, its time would literally come to a standstill.[12] This stretching of time sounds weirdly reminiscent of our forebears' elastic temporary hours and its warping by gravity seems an echo of the entrapment of time in nature's events. They are not the same, of course, any more than Democritus' atoms were the same as the atoms of today's nuclear physics, but nevertheless, the time of relativity is clearly something quite different from the consistently flowing absolute time measured by our clocks on earth.

Today, the relative time of the cosmos and the absolute time of earth run alongside each other, two different "times within Time." Because the effects of relativity on the surface of our earth are so miniscule and because we *experience* time and space as separate, our clock time surely never will be unseated by any sort of space-time clock. But it is conceivable that someday it may be replaced by something else. The temporary hours had a run of more than three millennia before the clock's equal hours took their place around seven hundred years ago. If and when our clock time is replaced, it won't be because time itself has changed, but only because we have found a new way to view it.

Chapter VII

The Mechanical Clock: The Machine

And even as wheels of clockwork so turn that the first,
to whoso noteth it, seemeth still, and the last to fly . . .
Dante Alighieri (1265–1321)

What was the state of technology late in the thirteenth century when the clock entered the world? Water-powered mills for grinding grain had existed since antiquity, but perhaps because of the availability of slaves, the ancient world never embraced mechanization with the eagerness seen in the Middle Ages. By the eleventh century, water power was also being used for fulling cloth, tanning leather, and forging iron. The windmill, invented in the twelfth century, spread rapidly, and by the fifteenth century the power of wind and water was grinding everything from paint pigments to pulp for paper to olives to mash for beer, as well as sawing, laundering, and operating the hammers of a forge.[1] It was a water- and wind-powered technology; the much more constant power of gravity had hardly been tapped. In terms of machine design, there were cranks, flywheels, and treadles, and the spinning wheel with its differential gearing had appeared by around 1280.[2] It was a time of exploration into the ways that the forces of nature could be harnessed by machines.

No machine reveals this spirit of exploration better than the clock. Although much about the mechanical clock was foreshadowed in the most sophisticated of the waterclocks, it was nevertheless a truly new, truly original machine. For all its ingenuity, the machine that first produced our clock time had just three essential components: a source of power, a regulator of that power, and a way to transmit the results to a face, or dial, where they could be translated into our culture's numerical language of time. (The earliest clocks lacked this last part—only the sound of bells told what time it was.) This deceptively simple description applies just as accurately to the atomic clocks of today as it did to the earliest iron behemoths. Through all the permutations and improvements of the past seven hundred years, this basic fact has never changed.

Into a world whose machines were powered by wind and water came the mechanical clock with gravity as its power source. For all its superiority to wind and water in terms of constancy, gravity had always presented a problem. Consider a weight on the end of a rope—the weight just keeps on going until it hits the ground. If the rope is wound around an axle, it will certainly cause a wheel to turn as long as the weight is in the air, but once it hits the ground, its power *to* power abruptly ceases. There may be constancy in gravity, but it would be very short-lived in such a machine. Gravity was powerful all right; the problem was figuring out how to control its power.

Although the mechanical clock's use of gravity as a power source was certainly innovative, its truly great invention was the device for controlling gravity's force: the escapement. All of the earlier timekeepers had used something that *flowed* to track the passage of time: water, sand, the burning of a candle, the slow flow of sunlight across a dial. The reasoning was that time, which seems to move continuously and evenly in the same direction, ought to be tracked by something else that does the same. However, it turns out that nothing *flows* at nearly a steady enough pace to precisely measure time. Surprisingly, time is best tracked not by something that flows but by something that oscillates at a regular interval and whose beats can be counted.[3]

This was the brilliant innovation embodied in the escapement, the regulator of the first clocks. The most widely used early escapement was known as the verge-and-foliot: the foliot was a horizontal bar on which were hung small weights; it was firmly attached to a vertical rod from which projected two pallets at right angles to each other (the verge). (In Italy, a balance wheel was used instead of the foliot.) The source of power for the clock was a saw-toothed crown wheel, which turned because a weighted rope was wound around its axle. As the crown wheel turns, it is stopped when it engages with the top pallet of the verge. This causes the foliot to oscillate, which turns the verge and allows the toothed wheel to escape from the top pallet, only to be caught again by the lower pallet. Again the foliot oscillates, causing the verge to turn and release the wheel from the bottom pallet. The wheel thus advances (or "escapes") one tooth at a time for each oscillation of the foliot. Although the speed of the crown wheel was primarily determined by the weight hanging from its axle, it could be secondarily regulated by moving the small weights along the foliot (closer to the verge meant a faster oscillation). On and on the foliot oscillates, causing the crown wheel to move slowly and steadily, tooth-by-tooth, as it is caught and released, caught and again released, by the two pallets. This is what makes the tick–tock of mechanical clocks.

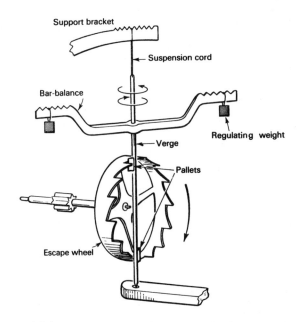

8. The verge-and-foliot escapement. The saw-toothed crown wheel (escape wheel) is turned by a weighted rope wound around its axle (not shown). As it turns, it is stopped by the pallets extending from the vertical shaft (verge). (In this drawing, the top pallet catches a tooth of the wheel.) This causes the horizontal foliot (bar balance) to oscillate, reversing the rotation of the verge so that the wheel is released by the upper pallet, only to be stopped again by the lower one. The process continues as the weight very slowly descends toward the ground. The driving weight, the length of the foliot's arms, and the position and size of its small weights all affect the rate at which the mechanism turns.

Thus, the escapement in these early clocks accomplished three crucial tasks: (1) it held back the fall of the weight so that the force of gravity would be stable and could continue to power the clock over a long period of time; (2) its oscillating foliot divided time into equal beats; and (3) its verge with pallets counted the beats. Clearly, all three tasks were very tightly bound together in these early clocks. Any tiny change in the foliot's oscillation or in the weight driving the crown wheel would trickle through the whole mechanism and cause the clock to become inaccurate. (The significance of this will become more obvious when we look at the pendulum in a later chapter.)

The last step that remained was to transfer the escapement's rationing of time to a dial or bell. The gear train (also controlled by the escapement) simply transmits the beats of time to the clock face. It is the beauty of the differential gearing of mechanical clocks and watches that thrills us, but such gearing had existed since antiquity and is not the heart of the clock. The gear train is like the ever-smaller blood vessels and capillaries of our bodies, but the escapement is like our hearts—it is the pump that rations out, beat-by-beat, the life force of the clock. This is all there was to it. In our digital wrist watches, the power source, regulator, and clock face have all changed, but in principle it is still the same machine.

Of all the machines that have transformed our world, we know the least about the origins of our clock. Although we will never be able to put them all together, let us look at some of the pieces of the puzzle we have been able to discover so far. A precious piece of evidence has been found in the lecture notes of an otherwise unknown Englishman named Robert. All we know about Robert is that he was in France around 1271 lecturing on Ptolemy's conception of an earth-centered cosmos at either the University of Paris or the University of Montpellier. Here is part of what was found in Robert's lecture notes:

> Nor is it possible for any clock to follow the judgment of astronomy with complete accuracy. Yet clockmakers are trying to make a wheel which will make one complete revolution for every one of the equinoctial cir-

cle, but they cannot quite perfect their work. But if they could, it would be a really accurate clock and worth more than the astrolabe or other astronomical instruments for reckoning the hours, if one knew how to do this according to the method aforesaid.[4]

We thus know that by 1271 people were either trying to come up with a clock or were trying to perfect one that already existed (both interpretations are possible from Robert's notes). Nevertheless, none of the earliest clocks have survived—the Salisbury Cathedral clock is the oldest to survive, but mention of it doesn't appear until 1386.[5] Since most early clocks were built of iron, this may seem surprising until we remember that they had a terrible breakdown rate. They were constantly being repaired, and we may assume that when one was irrevocably broken, its parts were salvaged and carted off to be used on another.[6]

We also don't know the name(s) of the clock's inventor(s). It seems probable that the clock was invented by monks, because of the temporal regimentation of monastic life, but whether or not this is true, it certainly owed its development to the Catholic Church. The mathematics required to construct gear trains required a high level of education, which was provided only by the Church during that period.[7] Also, the alarm mechanisms of the most sophisticated monastic water clocks were weight-driven devices that are the most likely forerunners of the clock's escapement.[8]

Our knowledge about the origins of the clock is also thwarted by the fact that only one word—*horologium*—existed for timekeeping devices at this time, and this word was used for *any* device that counted the hours. Thus, it is not possible to know with certainty whether a reference is to a water clock, mechanical clock, or even a sundial. Nevertheless, there is a huge jump in the number of records concerning *horologia* during the late thirteenth century. The account books of monasteries and cathedrals begin to mention costs of parts needed to repair clocks, as well as payments to *horologeurs* (clockmakers) to make the repairs. Although we cannot know for certain, it is reasonable to speculate that this spate of records refers to a new machine rather than to one that had been around for a long time.[9]

And, during the early fourteenth century, records begin to appear of clocks in cathedral towers that are almost surely mechanical: the first known public clock was erected in the tower of the Church of St. Eustorgio in Milan by 1309, and it was followed by others in cathedrals at Caen (1314), Norwich (1321), Florence (1325), London (1335), Milan (1335), Modena (1343), Padua (1344), Monza (1347), Strasbourg (1352), Genoa (1354), Bologna (1356), Siena (1359), and Ferrara (1362). For two astronomical clocks—that made by Richard of Wallingford (begun around 1330 and requiring thirty years to build) and that of Giovanni de' Dondi (completed in 1364 after sixteen years of work)—such detailed descriptions were left that precise working replicas have been built.[10]

The clocks in the cathedral towers were *huge*, and their erection was an arduous feat which often took several years. These giant wrought iron clocks could weigh as much as two tons, their bells four tons, and the weights that drove them as much as a thousand pounds. Building one clock required the combined efforts of up to a hundred workers: in addition to the clockmakers, there were blacksmiths, ropemakers, carpenters, bricklayers, plasterers, gilders, and bell founders. Once the bell and the huge iron clock were made, they had to be hoisted up thirty feet or more to the top of the tower. The fact that these great tower (turret) clocks were being erected in towns throughout Europe and the British Isles during the first half of the fourteenth century indicates that the machine itself must have been invented well before. Even if Robert's clockmakers had ironed out the wrinkles in their effort close to the time of his writing, this proliferation of huge working clocks within seventy-five years after its invention would still indicate an extraordinarily rapid technological advance as well as social acceptance of the new machine.

The astronomical clocks of Richard of Wallingford and Giovanni de' Dondi represent a different, and somewhat confusing, aspect of our story. As luck would have it, the two early clocks about which we have the most detailed knowl-

9. One of the countless fifteenth century cityscapes which prominently included a clock in a bell tower.

edge were not simply timepieces—these amazing machines were also mechanical models of the universe. The Dondi instrument, with its geared mechanisms for the movements of sun, moon, and planets and its perpetual calendar for all fixed and movable religious festivals, was really only incidentally a timepiece. It was a planetarium powered by clockwork. Our planetariums of today continue, in monumental form, this ancient tradition of which the Dondi clock was a medieval masterpiece.

This extremely long-lived tradition existed in China as well as the West. Although there were no mechanical clocks in the ancient world, there actually were geared mechanisms capable of reproducing the motions of the heavenly bodies in relation to one another. Like so much else in the ancient world, this tradition of planetaria was transmitted to the Islamic culture and from there made its way to the medieval West. The Wallingford and Dondi instruments are superbly complex examples of the tradition, but along with our own room-sized planetariums, the familiar moon dials on grandfather clocks also derive from this ancient desire to replicate the heavenly motions in an automated miniature. Of course, at this time the movements of the sun, moon, and planets were those of the earth-centered Ptolemaic world and not of our own sun-centered solar system.

So we have two separate instruments here: the clock for telling terrestrial time and the planetarium for duplicating the motions of the heavenly bodies. In the Wallingford and Dondi machines (and subsequently in thousands of others), the two were combined—the clock's mechanism powered the gears which made the planets move. The question is: How did they become combined? Did the clock originate separately, as we have suggested, or did it somehow grow out of the older tradition? This has been a much debated question, but most students of the subject now believe that the clock was indeed invented independently of the planetaria. As evidence, we have the monks' intense interest in earthly timekeeping, we have the words of Robert the Englishman, and we have what were almost surely working tower clocks long before the Dondi planetarium or the Wallingford astronomical clock.

We also have indication of a new profession (that of clockmaker) which usually comes along with the invention of a new technology. Robert mentions clockmakers in 1271, and they also show up as the payees in cathedral account books. In addition, Giovanni de' Dondi spent 130,000 words and 180 diagrams describing his masterpiece—his attention to detail ran to the length of the studs and precisely where to drill holes.[11] This was clearly not a man afraid of saying too much or of insulting the intelligence of his reader. And yet, when it came to describing the clock part of his magnificent apparatus, he simply referred to it as a common clock and said the following: ". . . if the student of the ms. cannot complete this clock for himself he is wasting his time in studying the ms. further."[12] What we would have loved to have known was considered too common knowledge for even the conscientious Dondi to spell out. And there are other words to show that the ordinary time-telling clock was something completely distinct from planetaria. Petrarch describes his friend Dondi's instrument as something "which the uneducated people think is a clock . . ."[13]

Far from being an offshoot of the planetaria, everything we know today indicates that the mechanical clock was invented for its own sake. Only later were the two combined: either the phases of the moon and zodiacal indications were added to clocks as embellishments, or, as in Dondi's masterpiece, clockwork was used to power a microcosm of the Ptolemaic solar system.[14]

By connecting earthly time with the movements of the other planets, these astronomical clocks highlight two crucial facts about our clock time. First, in using our clocks to measure the tiny bits and pieces of our days, we often forget that the twenty-four hours of our earth's rotation belong in this more cosmic context. After all, it was the apparent motions of the heavenly bodies that first got humankind to thinking about the possibility of measuring time. But it is also of significance that it is the earth-centered Ptolemaic concept which is pictured in these astronomical clocks, and that in terms of timekeeping, *it does not matter at all*. The twenty-four hours of the first clocks and our own are identical. In order to measure earthly time, it makes no difference whether the sky revolves about a stationary earth or the earth's rotation causes the sky to appear to move.

Chapter VIII

The Early Machines

These great clocks were, like computers today,
the technological sensation of their time.

David S. Landes

From the time the clock was invented until now, improvements upon it have never ceased. People accepted the first clocks eagerly in the late thirteenth century, and they continue to embrace its new forms today. If solutions to difficult horological problems were sometimes slow to come, clockmakers hung on like terriers until they finally figured them out. In terms of transformations in the physical mechanism itself, the size of the time periods it registers, its accuracy, and the degree to which clocks impinge on the lives of the people they "serve," the movement has unfalteringly been in the direction of more and better.

Aside from technical difficulties along the way, the only real obstacle to the clock was the Church with its focus on eternity and its division of the day into the canonical hours of worship. It is a strange twist of fate that the mechanical clock, which in all probability was invented by monks, should now rise up and, so to speak, bite the hand that fed it. For the clock carried in its wake a faster-paced and more worldly way of life that was antithetical not only to the pious discipline of the monks but to the life the Church desired for its parishioners. Nevertheless, as it did on the usury question, the Church slowly accepted the need its congregants felt for the new time, and the manner of its acquiescence was to allow mechanical clocks to ring the twenty-four equal hours from its towers. Although this didn't happen overnight, a major issue was never made of it and it certainly occurred far more rapidly and with far greater ease than was the case with the issue of usury. Already in 1336, the first clock specifically recorded as striking all twenty-four equal hours was in the tower of St. Gothard's church in Milan.[1]

In 1370, King Charles V of France finally got fed up with the cacophany and confusion created by the bells of Paris ringing at so many odd times that they almost made it worse than having no bells at all. Some bells rang the canonical hours, others rang the hours for the beginning and ending of the work days of various professions, and yet others rang their approximation of the current equinoctial hour. Charles decreed that all of the bells in Paris must ring simultaneously with the clock on the Royal Palace, and since it struck the twenty-four equal hours, Charles' edict helped to undermine the dominance of the Church's canonical hours. "Church towers, built to salute God and to mark man's heavenward aspirations, now became clock towers."[2] The trend continued, as can be seen (and heard) by the countless old equal-hour clocks on church towers throughout Europe and the British Isles.

To see how different it might have been one has only to look at the Greek Orthodox Church, which did not allow mechanical clocks into its churches until this century. The blasphemy that the mechanical clock represented for the Greek Orthodox Church was not simply its departure from tradition; the clock was blasphemous because "the mathematical division of time into hours, minutes and seconds had no relationship with the eternity of time."[3] To use a machine to divide time into minute portions was necessarily to shift humankind's focus of attention away from eternity and toward the immediate things of the here-and-now. We think of time as something real but not as having any meaning, but in fact the meaning that our particular culture has given it is so deep in us we are unable to see it. This great awakening to earthly life is usually associated with the Renaissance, but with the mechanical clock it had already been initiated in the last decades of the thirteenth century.

Certainly the move to equal hours was gradual throughout the fourteenth century and even the fifteenth, and it coincided with the actual presence of a tower clock, which meant that it gained ascendancy in the towns and cities. In the countryside, and thus for the vast majority of people, time continued to be marked by such natural "clocks" as the rooster's crow at dawn and the rising and setting of the sun itself. In vernacular French literature, time was still given in canonical hours throughout the fourteenth century,[4] and the great popularity of Books of Hours continued into the sixteenth. Christopher Columbus, traveling in 1492 on the ocean waves where no weight-driven mechanism could follow, measured time by the canonical hours. Perhaps at no other moment in history did the various "times within Time" so jostle one another in everyday life. Today, our vast scheme of civil chronology and our knowledge both of geological time and of time's relativity may give us a wider range of ways to think about time, but in everyday life our clock time is universal. The seasonal rhythms and annual round of religious holidays are leitmotifs which we cherish, but the clock is king in the day-to-day. If we also used other systems to mark daily time, we could more easily see that our clock time is a specific cultural invention. When we say that it is three "o'clock," we are using a shortened version of "of the clock," begun in the Middle Ages when it was still necessary to designate the mode of timekeeping. It is the very dominance of the clock today that has caused us to confuse *its* time with time itself.

So how did we get to here from there? Granted, seven hundred years is a long time, but it is also a long, long journey from huge iron bells tolling only the moment of the hour change for everyone in town to our individual wrist watches continuously and inescapably ticking off each second. Although we carry with us an abstract concept of time bred from the mathematical divisions of our clock time, that concept is bound to the machine itself. If my watch breaks, I am incapable of intuiting the time from my abstract concept. I simply do not know it, and have to inquire of someone who is in possession of a watch. Our story must thus be of the machine itself and how it developed and multiplied.

Because the huge turret clocks made the greatest impact on society and because our earliest reliable data refer to them, we speak of them as the first. It is probable that the *very first* mechanical clocks were small chamber clocks, which were replicated in large in the towers.[5] Assuming that the first clocks developed from small alarm mechanisms, there almost necessarily would have been smaller precursors to the giant constructions in the towers. We know for certain that small clocks begin to appear in various inventories by the middle of the fourteenth century. Charles V's horological interest extended beyond the synchronization of the public clocks of Paris, because two chamber clocks are mentioned as among his possessions upon his death in 1380. One of these was an *orlage portative*, indicating that it was small enough to be moved. Not surprisingly, the inventories that have survived are mostly of royal or princely belongings, so that we cannot know how many other chamber clocks were made during the fourteenth century. However, references to clockmakers continue throughout the century, which gives us reason to infer the existence of many more domestic clocks than are mentioned in royal inventories.[6] These clocks, whether fixed or small enough to be portable, would all have been weight-driven—they were essentially miniatures of the tower clocks, but made of brass or silver instead of iron.

Except for the invention of the clock face, or dial, by Giovanni's father, Jacopo de' Dondi, in 1344, there were few innovations in clock design during the remainder of the fourteenth century. Some clocks began to strike on the quarter hour, but until the middle of the seventeenth century, most clocks had only one hand and the dial of most public clocks was still divided only into hours and quarter hours.[7] When we say that it is half past eight or a quarter to nine, our language harks back to these old clocks. The slowness to put numbers on public clocks may have been due to the fact that most people were not only illiterate but innumerate; they would not have been able to read them. Of course, given that the worst of these early clocks lost as much as an hour a day and good ones a mere fifteen minutes, the absence of numbers on the face was no great tragedy. This failure of accuracy was due to friction among the parts, that all gears were cut by hand and, above all, that the time divider (foliot or balance wheel) had no natural period of its own. This last factor is a pivot upon which our story hinges, and we will return to it in the next chapter.

Although many expressions of irritation about the clock's lack of precision have come down to us, this in no way dissuaded people from wanting them, accurate or not. Speaking broadly, the increasing impingement of clock time upon us is simply a matter of there being more and more of them, and of their getting smaller and smaller and closer and closer to us. The weight-driven clock presented many obstacles to the development of greater miniaturization and portability. In the first place, it had the problem of having to be hung high enough so that its weights could fall for awhile before it had to be rewound. Another problem of weight-driven machines is that they have to be vertical and stationary in the earth's gravitational field in order to work. A weight-driven clock certainly wouldn't work on a ship being tossed by waves, and it also wouldn't work on a pocket watch being moved about and turned sideways by its wearer. Anything weight-driven also has a downward limit in terms of size—even if it were stationary, a watch is just too small to work by means of weights.

A solution to the problem came very quickly. We don't know who first realized that the tendency of a tightly coiled spring to unwind could serve as an alternative to weights as a source of power in clocks, but such a spring (called a mainspring) began to be used sometime early in the fifteenth century. Unlike weights, a spring has no dependence on gravity and unwinds equally well in any position; it also works just as well small as large. The mainspring was the first step on the long road that would lead to the ultimate portability not only of our wrist watches but also of the timekeepers used for navigation on our space travels. There was just one little problem. With the weight-driven clock, each drop of the weights was so short that acceleration didn't enter the picture; the wheel train always received an equal force whether the weights were near the beginning or end of their fall. The same cannot be said of an uncoiling spring, which imparts greater force the more tightly it is coiled; as it unwinds, its force diminishes. If a way could not be found to keep the force equal, all the other virtues of the mainspring would be for nought.

The solution was one of the most ingenious devices in a technology filled with ingenious devices: the fusee. This was simply a cone-shaped piece of brass which was attached to the mainspring by a cord or chain, so that the force extended to the wheel train resulted from the combined effect of the fusee and the mainspring. When the clock is run down, i.e., when the spring is uncoiled, the cord is wound around the barrel containing the mainspring. As the fusee turns, it winds the cord about it, starting from the wide end of the cone and moving up to the narrow. In the process, it pulls on the spring and winds it up again. Then the process begins all over again. When the spring is tightly-coiled, it unwinds the cord from the narrow end of the fusee, where its radius and leverage are small. As the spring runs down, the cord unwinds from ever wider parts of the fusee, where the leverage increases. In this way, the fusee compensates for the diminishing force of the uncoiling spring, and the two together deliver a constant power to the wheel train.[8]

As early as 1430, a clock with a mainspring and fusee is recorded as belonging to Philip the Good, Duke of Burgundy,

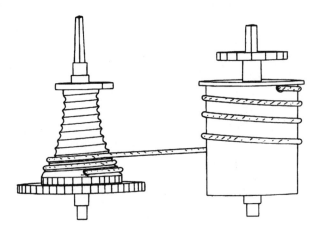

10. Fusee. The mainspring is coiled within the barrel (right) and the cord is attached to it. When the spring is wound tightly, and its power thus greatest, it pulls the cord from the narrow top of the fusee, where leverage is minimal. In the drawing, the spring is nearly unwound and its power lessened, but the greater leverage at the wide end of the fusee equalizes the motive power of the spring. Thus, the clock is prevented from gaining time in the first hours after winding and from losing time as the spring unwinds.

and records of more spring-driven clocks appear throughout the remainder of the century. With this new technology, the way was opened for production of smaller and smaller domestic clocks, and from them it was a quite natural step to watches that could be worn on the person. The legend giving Peter Henlein of Nuremberg credit for inventing the watch early in the sixteenth century has now been discredited, and it appears more likely that it was invented in Italy during the late 1400s.[9] We certainly know that Francis I of France paid for tiny watches to be placed on the hilts of two daggers in 1518, and that Henry VIII presented a gold watch to his ill-fated fifth wife, Catherine Howard, in 1540. These early watches were far thicker than anything we would recognize as a watch. They were status symbols and pieces of jewelry as much as timekeepers. They had thick, ornately decorated cases, but kept time so poorly that a minute hand, much less a second hand, would have been pointless.

Nevertheless, these spring-powered watches and small clocks represented a new stage (plateau, really, as we will soon see) in the evolution of the mechanical clock. Bells ringing the equal hours from church towers were a very intermittent and momentary reminder of the time. For fifty-nine minutes out of sixty (if minutes had been counted), people were left to their own subjectivity for their sense of time. For the vast majority of people, this intermittent and momentary reminder of clock time was all there was, but watches and domestic clocks gave to the few who could afford them a wholly different awareness of time. Unlike the tower clock, these machines were always visible, and even before the minute hand arrived they enforced a sense of the ever-presence of clock time and brought in their wake all our thrifty habits of monitoring time so as to use rather than waste it.[10]

Not only did the earliest watches inaugurate the sense of continuous mechanical time which is still ours, but they finally got the new profession of clockmaker off the ground. In the beginning, there had not been enough demand for the product to make clockmaking a full-time job. The builders of the big iron tower clocks were recruited from other metalworking crafts: blacksmiths, founders, locksmiths, and makers of cross-bows and bombards. They were local men pulled out of their usual professions to help construct the town's big clock. When it was finished, they went back to their old jobs. Only the best of them mastered the many facets of clockmaking and became full-time, and even this was possible only if they moved around from town to town, following available work. There are many accounts of princes and dukes searching high and low for someone capable of repairing their clock or building a new one.

This situation changed once watches came on the scene. The market broadened to include wealthy bourgeois and, for the best artisans, clockmaking became a full-time job in one place. Although the profession had always required versatility in its workers, at the same time it had always been a collaborative effort by members of specialized crafts. With the watch, the trend went in the direction of specialization. The tower clockmakers and those who worked on domestic timepieces split off from one another, and eventually the clockmakers and watchmakers did also. Since watches in particular were as much ornaments as machines, a further division was made between the artists who made the outside (the decorative case and dial) and the engineers of the clockwork itself. The making of the spring was such an art that it was the one thing that even the most versatile master clockmaker always contracted out.[11]

So, by the early seventeenth century there was a well-established profession churning out timepieces for an eager and ever-broadening clientele, and *still* the instruments didn't keep good time. A plateau had been reached, above which was a ceiling that could not be broken through. On the outside, clocks and watches became more and more beautiful, but on the inside their mechanisms kept time only a little better than had the earliest turret clocks. The state of affairs is perhaps most succinctly revealed by the answer to the following question: What was the ultimate standard against which these clocks were checked? People knew that they were inaccurate because they all rang at different times, but how on earth did they determine the *right* time? It couldn't have been

11. Woodcut of a fifteenth century clockmaker using a quadrant to measure the sun's altitude so that he can regulate his clocks by the sun. At this time, many clockmakers also made their own astronomical instruments.

from any of the clocks. It was from an old friend, a friend whom we thought the clock had long ago sent into obsolescence. It was the *sundial!*

This was not the temporary-hour sundial of the ancient and medieval worlds. During the Renaissance a way had been found to make a dial that showed equal hours. It was in fact this need of the new mechanical clocks for an outside standard of accuracy that had spurred the effort to come up with the equal-hour dial. Sundials underwent a rebirth of their own, and by the seventeenth century a special branch of education had grown up that was devoted to the science of "dialing." If proof were needed that sundials served as the ultimate time reference for mechanical timepieces, it can be found in the many watches that came with a sundial inside the cover, complete with a compass to align the gnomon in a north/south direction.

If the very best sundials under the clearest skies can't show time to less than a five minute period, and if they were the final arbiters, what hope was there for improvement in mechanical timekeeping? A ceiling had indeed been hit. Always in the strange and paradoxical story of time we keep being thrown back on nature, and this return of the sundial is a case in point. As we will see in the next chapter, it would take another of nature's gifts to bring the clock through the ceiling and into a new realm of accuracy.

Chapter IX

The Pendulum

The universe "is like a rare clock . . . where all things are so skilfully contrived, that the engine being once set a-moving, all things proceed according to the artificer's first design. . . ."

Robert Boyle (1627–1691)

While attending prayers at the Cathedral of Pisa in 1583, the nineteen-year-old Galileo Galilei (1564–1642) got sidetracked by the swinging altar lamp. The longer he looked at it, the more it seemed that the swing took the same amount of time whether it made a wide arc or a tiny one. He used his own pulse to time the various swings of the lamp, and by the time prayers were over, the principle of isochronism had been discovered: the length of time (period) of a pendulum's swing depends on how long the pendulum is and not on the width of its swing. The explanation for Galileo's discovery is really quite simple: when a pendulum swings in a wide arc, the object on its end (lamp, ball, whatever) falls with greater velocity, and the longer distance it has to travel is made up for by its increased speed.

Whether or not the story about Galileo's distraction from prayers is true, he surely made his great discovery in a moment—the answer must have come almost as quickly as the question. This brief moment in a great mind has had an immeasurable impact on all our lives, because what Galileo had discovered was a natural periodicity. At the large end of the scale, the twenty-four-hour day and the 365.2422-day year are natural periodicities; at the small end, the vibrations of tuning forks, quartz crystals, and cesium atoms. Even the human pulse, which Galileo used to make his great discovery of the pendulum, is a natural periodicity. The world is full of all kinds of natural periodicities and some of them are extremely precise. Mankind cannot jostle them loose. They are built into the universe. Now, if one of these natural periodicities could be made to control a clock, it would run quite effortlessly and with a degree of precision far beyond that previously achieved with much greater effort.

Why couldn't the early clocks keep better time? In addition to errors caused by friction and the fact that most gears were cut by hand until the seventeenth century, the main problem lay with the regulator. Two functions which should have been separate were combined in the early regulators: the oscillating foliot (or balance wheel) to *determine* the segments into which time would be broken, and the verge with its pallets to *count* these segments by blocking and releasing the crown wheel. In the early clocks, these functions were linked together so tightly that the foliot's beat was necessarily affected by fluctuations in the force transmitted by the crown wheel. When the frequency of the beat depended on the interplay of so many different mechanical factors, it is no wonder that timekeeping was so chancy. What was needed was something whose frequency was independent of the mechanical parts of the clock, something with a reliable frequency inherent in itself that could somehow be used to control the time-making of the clock.[1] Such a device was of course the pendulum.

Although he never created a working clock, Galileo clearly realized this was an ideal use for the pendulum. In his old age, he came up with a design for a pendulum clock, and his son Vincenzio was in the process of constructing this clock according to his father's design when he died in 1649. It was the Dutch scientist Christiaan Huygens (1629–1695) who at the age of twenty-seven devised the first pendulum clock that was ever completed. Together with Salomon Coster, he built it in The Hague in 1657. The pendulum clock was accepted everywhere with the same eagerness that had greeted the original machine itself.

If clocks were to be successfully regulated by one of nature's laws, then a sort of sanctuary for the free play of nature would have to be created within the patently unnatural confines of the clock's apparatus. The power from the weight-driven wheel train had to reach the pendulum in order to keep it going, but if it were kept as constant and slight as possible and nothing else in the clock affected the pendulum, it would be left free to tick off equal bits of time in obedience to the isochronism built into it by nature.

This is where the brilliant idea of the anchor escapement came in. Huygens' clock still used the same verge escapement that had been paired with either the foliot or balance wheel

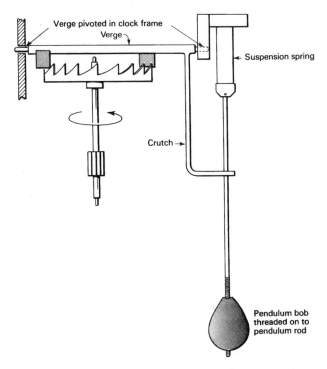

12. Pendulum clock. The natural oscillation of the pendulum has replaced the mechanical oscillation of the old horizontal foliot with its weights. Otherwise, the instrument is essentially the same. Now it is the beat of the pendulum that controls the speed at which the verge with its pallets turns. As the pallets catch and release each saw tooth of the crown wheel, the pendulum's beats are counted, and the count is subsequently transmitted by the gear train (not shown) to the clock face.

for the past four hundred years—except that in his clock the pendulum took the place of the foliot/balance wheel as the mechanism for determining the beats of time. Just as these earlier oscillators had had to make wide swings in order to turn the verge so that its upper and lower pallets could alternately stop and start the escape wheel, so too Huygens' pendulum had to swing in an arc as wide as twenty degrees or more in order to turn the verge.[2]

As with so many horological inventions, there is dispute about the authorship of the anchor escapement. Whether

Robert Hooke or the anchorsmith William Clement or someone else was first, we know that a clock with this new escapement was built by Clement in 1671 for King's College of
Cambridge University. The beauty of the anchor escapement
was two-fold: not only did its streamlined form minimize the
interference of the rest of the clock with the pendulum, but it
also allowed the pendulum to make an arc of only three or
four degrees. Instead of having to turn a verge, the pendulum
simply rocks the anchor as it swings. Tooth-by-tooth, the escape wheel is caught and released, caught and released by the
tongs of the anchor. The principle is the same as in the earlier
mechanisms, except that now, with the pendulum's natural
periodicity governing the movement, the escape wheel turns
at a nearly constant rate, sending a far more accurate beat of
time along the gear train to the hand we see moving around
the dial.

13. The anchor escapement. The pendulum (not shown) is attached to the
anchor, which has replaced the verge with its pallets. As the pendulum
swings, it simply rocks the anchor so that it catches and releases each tooth
of the crown wheel. There is now far less opportunity for the clock's mechanism to adulterate the natural regularity of the pendulum's beat, and it can
be more directly and accurately transmitted to the clock's face.

Already in 1644, the French mathematician Marin Mersenne had discovered that at the forty-five-degree latitude, a 39.1-inch-long pendulum would complete its swing in precisely one second.[3] When pendulums of this length were put into clocks, it became a simple matter to transmit this count to the clock face—the passage of sixty seconds each minute was shown on its own little dial just below the "12" on the larger dial. Clocks with anchor escapements and seconds-beating pendulums (known as Royal pendulums because of their superb performance) varied by no more than ten seconds per day. This was nearly two orders of magnitude less than fifteen minutes per day, which was the best that most verge-and-foliot clocks could manage, and considerably better than the short pendulum and verge types of Huygens' invention.

Contrary to popular belief, the tall clocks known as grandfather clocks did not come into existence specifically to accommodate the 39.1-inch pendulum. They first appeared around 1660, just after the Restoration of the English monarchy, at a time when the enforced puritanism of Cromwell's England was giving way and furnishings were again becoming more elaborate. The tall trunks of these first grandfather clocks were simply a perch for clocks so that their weights could fall for a long time before hitting the ground. Once the Royal pendulum was invented, the tall case of course became an ideal structure in which to house it, along with the weights.[4]

Let us return for the last time to the sundial and see how its status as ultimate time standard was affected by this exponential increase in the clock's accuracy. With only a ten second per day variation, many different pendulum clocks were able to differ among themselves and yet all sound within the five-minute range of a single sundial reading. In other words, the pendulum clock utterly destroyed the sundial's validity as a time standard. But it destroyed it in yet another way: by helping to prove that the sun's speed across the sky differs from one day to the next, so that a more precise sundial wouldn't have done any good anyway. The earth moves in an elliptical rather than a circular orbit around the sun, and is actually closer to the sun by over three million miles in January than it is during July. If we recall Kepler's Third Law requiring that the closer a planet is to the sun, the faster it moves, we will realize that every day in January the earth moves faster about the sun than it does in July. Add to this the fact that as the earth circles the sun, it spins on its axis at a $23^1/_2$-degree angle from the equator, and you have the two factors whose combined effects cause the apparent change in speed of the sun across the sky. When compared to the time on a pendulum clock, the sun may appear to travel as much as sixteen minutes ahead or as much as fourteen minutes behind; on only four days during the year do the two coincide. The difference between the natural time of the sun and the uniform, mean time of the mechanical clock is known as the equation of time. Conversion tables were created that charted the difference between the two times for every single day of the year, and these tables were pasted on the insides of clock cases.

The conversion tables represent a crossroads. Which time was to be considered "real?" Because setting the new clocks to sun time, as had always been done in the past, would now have meant resetting them with great frequency, most conversion tables were used to set clocks to the correct mean time. When the clock ran down, a sundial could be read, the comparable mean time found on the conversion table, and the clock could be set according to mean time. But the conversion tables could also be read the other way in order to find out the

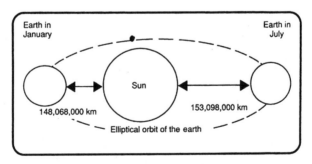

14. The earth moves fastest in its orbit each day in January, when it is closest to the sun, and slowest in July, when it is farthest away.

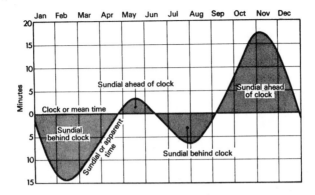

15. The equation of time.

"real" time as determined by the sun. Indeed, extremely complex and ingenious equation clocks, which showed both sun time and mean time, were first invented in 1660 and continued to be built throughout the eighteenth century. These clocks appear not to have been made for the purpose of verifying mean time, but rather for converting it into the "real" sun time by which the vast majority of people still lived.[5]

Eventually, and inevitably, a role reversal took place between the mechanical clock and the sundial. To see how decisively the definition of real time was reversed, one needs only to look at a common type of modern sundial which contains conversion tables so that the sundial's reading can be corrected to clock time. Such sundials came into being very soon after the conversion tables were created, and with them the sundial was demoted from being the regulator *of* the clock to being regulated *by* the clock.[6] The mechanical clock had finally won, and timekeeping had been lifted and liberated from the sun's yellow rays. It was now fitted to the rotation of the earth, and not to the reflection of that rotation in the apparent motion of the sun. With time no longer told by the sun's rays, the service of the sundial wound down. And yet, it is the oldest known scientific instrument, and it continues to be used today. About what other instrument can it be said that it is both the earliest of its kind and still in use today?

Chapter X

The Longitude:
Where Where Becomes When

Gulliver believes that only the immortal struldbrugs
*have a chance to "... see the discovery of the
longitude ... and many other great inventions
brought to the utmost perfection."*
Jonathan Swift
Gulliver's Travels, 1727

Although we have been to the moon and have done unmanned experiments on several of the other planets, in terms of clock time even our own solar system belongs to the realm of the great beyond. Based as it is, and must be, on the rotation of our planet, our clock time necessarily applies only to the surface of the earth. With their different rotation speeds, it wouldn't make sense on any of the other planets. It also wouldn't make sense because, as Einstein has shown, the gravitational fields of all celestial bodies affect time—a clock would run slower on Jupiter because its mass is 318 times greater than the mass of earth.[1] In relation to our own solar system—not to mention the rest of the universe—our clock time is thus as irremediably local as is the sundial to a particular latitude, but within the confines of the earth it has a universality which makes it utterly different from the sundial. Unlike the sundial, the clock works equally well anywhere, and with the radically increased accuracy brought by the pendulum it could be used to coordinate the time everywhere on the globe. But first many other things had to be discovered about time on earth, and a *need* for such world-wide coordination had to arise.

We tend to imagine time after the model we use for history or for the geological time scale. We see it as something moving straight ahead in one direction, from the past into the future. It feels as if it belongs to its own dimension, entirely unrelated to space. Our twenty-four time zones circling the globe and causing us to change our watches every fifteen degrees of longitude don't quite fit with this image. They make our earth seem a kind of whirling dervish of a time machine, mixing time with space as it makes its way from past to future after the fashion of a ball rolling down a hill.

A time machine it truly is. And if it does not conform to the simple dead-ahead image of time we carry in our heads, earth time nevertheless gets us from the past to the future very well indeed. The question is: how did humankind discover the riddle of time and our turning planet? And, once discovered, how did we rein in all the far corners of the earth so that they became part of a single entity in terms of time? The answer is best seen as a giant puzzle, the pieces of which were slowly found and pieced together. As in a puzzle, some pieces were discovered very early and some only at the last minute—there are pieces of the universal earth time puzzle which are more than two millennia old, and others that are only a century. But however various the pieces, the motivation for putting the puzzle together was always the same: the need of people to travel about the earth.

When we talk about travel, aren't we talking about space rather than time? One of the great chapters in the history of human invention—the story of finding the longitude at sea—will bring us the answer, but first we must make another detour from our subject and take a brief look at the events which led to the critical need to find the longitude at sea, because it was on the water rather than on land that the connection between travel and time first showed itself to be of perilous consequence.

Neither the vast expanse of the Roman Empire nor the lands encompassed by Europe's burgeoning trade during the late Middle Ages covered anything more than a small part of our planet. In terms of travel, we are only speaking of the lands that surround the Mediterranean and the other nearby inland seas, the British Isles, and the parts of northern Europe and Scandinavia facing the Baltic Sea. Three-quarters of the world simply did not exist in this story: all but the topmost part of Africa, the Atlantic and Pacific oceans, the whole American continent, Australia, and Antarctica. Although there was trade with Asia, very few Europeans had traveled to it, and its geography was only dimly perceived. In the West, the world stopped at the Atlantic ocean.

Travel by sea in this small world was a fairly simple matter. Ships could follow the coastlines of the various seas and have no fear of getting lost. Charts called portolans (harbor guides) delineated all the vital miscellany of facts about the

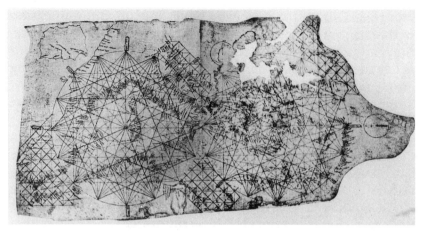

16. The "Carta Pisana," c. 1300 A.D. The oldest surviving portolan chart. The lines which crisscross the chart indicate the directions of prevailing winds or compass points; the ship's pilot used them to plot his course from one harbor to another.

coastlines: the depth of the water, direction of the winds, treacherous rocks and shoals, safe harbors, special landmarks, and any other information which could help to guide a sailor. These charts have been found from the fourteenth century onward, and similar navigation aids are known to have existed in the ancient world. When Prince Henry the Navigator began to send Portuguese vessels down the west coast of Africa during the 1420s, they were soon in strange waters for which no comforting portolans existed. The waters became even stranger at the end of the century, when Vasco de Gama rounded the Cape of Good Hope and crept up the east coast of Africa until he sailed out into the Indian Ocean and safely arrived in Calicut near the southern tip of India in 1498. If the Indian Ocean was a great emptiness, how much more so had the Atlantic been for Columbus just six years earlier?

On a tiny boat, and with no outside communication, to endure the vast emptiness of the ocean for weeks on end is something we cannot conceive of. Add to this a great uncertainty about ever reaching land, and the situation is terrifying. However suspenseful our trips to the moon, they were nothing like this. Our astronauts were never out of communication with earth, and they always knew exactly where they were; they never had to fear getting lost. They knew they might not reach the moon, but there was no chance of their bumbling onto another celestial body (a New World) along the way. Space is already known to us, mapped out for us in a way that the surface of the earth was not even five hundred years ago.

What caused Columbus to set out into the Western Ocean Sea, and why was he lost in a way that our astronauts can never be? Incredibly, the answer to both questions involves the work of a man who lived in the second century A.D.: Claudius Ptolemy. Today we know Ptolemy primarily by his earth-centered conception of the universe. What is less well known is that Ptolemy also wrote eight books on geography. These geographical writings were lost for over a millennium, and re-entered the culture of the West, as did so many other great relics of the ancient world, by way of the Arabs. Written in the Greek language, the manuscripts first reappeared in the early thirteenth century. They then had to be translated into Latin, but by the early fifteenth century many copies were circulating around Europe. We know that Columbus owned one. Along with the manuscripts went twenty-seven maps, which were certainly rendered according to Ptolemy's cartographic ideas although they may not actually have been made by him.

Of the many errors contained in these maps, two served as a catalyst for an experiment which was the first step toward giving us the global world we take for granted today. Ptolemy's first error was to estimate the circumference of the earth at only three-quarters of the real figure, and his second was to extend Asia and India (the Indies) much farther eastward than their real boundaries. These two errors combined to give Columbus the reasonable idea that a relatively short voyage west across the Atlantic would bring him to the Indies and to their spices and silks so craved by Europeans. The names given to the West Indies and to our American Indians reveal both Columbus' desired destination and his reason for believing he had reached it when he landed in the Bahamas.

(The distance to these islands was very close to that which Columbus had calculated would bring him to the Indies.) Remember that there was no knowledge to contradict these maps, and Ptolemy's name was an imprimatur; Columbus made his voyage in what was still an earth-centered Ptolemaic universe. Had it not been for these errors in Ptolemy's maps, there is no telling how much longer the New World of the American continent might have gone undiscovered.

For all the theories about how history happens, much of it comes down to plain chance. The ancient world gave birth to many great geometers, but it was Ptolemy's writings which came down to us first and they became the gospel to which the Renaissance willingly deferred. Ptolemy had rejected the theory of a sun-centered cosmos developed by the Greek astronomer Aristarchus, and he had also rejected the very nearly accurate estimate of the earth's circumference made by another Greek, Eratosthenes. (It is impossible for us to conceive what our world would be like today if these two facts, both determined in the third century B.C., had been accepted and built upon from the time of their discovery, or even if they had been accepted by Ptolemy four hundred years later.) But in another case Ptolemy chose rightly, and from the moment his works were rediscovered until today this choice has remained (literally) the framework upon which the particular shapes of our earth's surface are hung. In the place of nature's wind and compass directions, which carve up the surface of the portolan charts like a kaleidoscope, Ptolemy's maps contain the calm, abstract boxes made by latitude and longitude lines.

Despite their virtue of an empirical approach, the portolans told local, piecemeal stories—not only did each tell the tale of a tiny part of the world, but no two portolans of the same place were identical. Without some abstract system of coordinates which could be applied to all of them, they *couldn't* be identical.[2] The relationship between the local, worm's eye view of the portolan charts and the eagle's eye image of the whole globe achieved with the abstract grid of latitude and longitude lines is analogous to the one we have already seen between the sundial and the pendulum clock in the realm of time. Just as our zone time today is a man-made, mathematical grid fitting over the natural fact of the rotation time of our planet, so the latitude and longitude lines are an equally man-made mathematical grid fitting around the skin of the earth's surface. Both grids stand apart from the natural rhythms to impose an abstract form upon the the content of nature, paradoxically allowing us not just to gain control over nature but to comprehend it more truly.

By the third century B.C., the Greeks had used astronomical calculations to devise the first three reference lines circling the earth: the equator and the tropics of Cancer and Capricorn. Eratosthenes then added more east–west latitude lines, placing them so that they ran through familiar places.

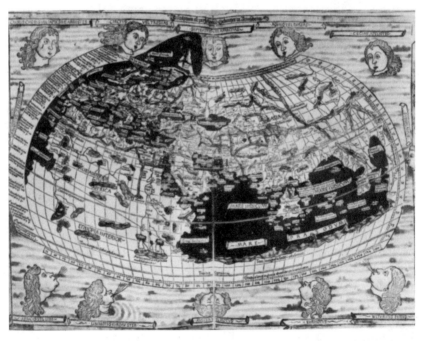

17. World map from the Latin edition of Ptolemy's *Geographia*, printed in Ulm, Germany, in 1482. The direction of the winds, which was an integral part of the portolan charts, has now been relegated to decoration: the heads surrounding the map merely personify the winds. The map itself derives its structure from the abstract grid of latitude and longitude lines.

A century later Hipparchus rationalized these lines, insisting that they be truly parallel, obedient only to the geometry of the planet and not to the important places of man. He also devised a system of longitude lines encompassing 360 degrees and running north–south from pole to pole. Ptolemy accepted this system of coordinates and improved and codified it. He also required that maps be drawn to scale, and he subdivided the 360 degrees of both latitude and longitude lines into smaller segments: sixty minutes to a degree and sixty seconds to a minute.[3] (Both our clock's circular face and the sphere of our planet owe their divisions to the four millennia-old sexagesimal system of the Babylonians.) Thus, when we look at the delicate latticework traced out on any globe today, we are looking at a double gift from the past—not only the Babylonian number system but also a scheme for delineating our planet which was developed over centuries in the ancient world and came down to us whole in the work of Ptolemy.

Although the ancients invented an abstract form for our globe, they knew only a tiny part of its true content. It would take the great explorers of the age of discovery to bring into view that gorgeous content, and fill up the abstract squares with new oceans, continents, islands, rivers, and mountains. This they did with extraordinary rapidity. Amerigo Vespucci discovered South America in 1499; Balboa sailed into the Pacific Ocean in 1513; and in 1522, just three decades after Columbus' journey, the survivors of Magellan's great voyage had made it around the world. We now had both form and content, but how did they get matched up? Slowly the new discoveries were grafted onto Ptolemy's maps and then slowly the fictive portions of his geography began to fall away. Although the new maps were surprisingly good considering the navigational and cartographic tools available, the shapes and sizes and placements of the new discoveries had none of the exactitude we take for granted in our maps today. The American continents first appeared on Martin Waldseemüller's map of 1507, but had they not been labeled as such no modern person could guess their identity.

This brings us back to our second question concerning Columbus: why was he lost in a way that our modern navigators, on earth and even in space, can never be? If you can't say exactly where you are, then you can't make a map to get you back again to precisely the same place. But if you can do the one, you can do the other—navigation and cartography are two sides of the same coin. Despite his superb navigational skills, Columbus could do neither, and thus he was, technically speaking, lost. What he could not do was determine precisely where he was in relation to the abstract grid of latitude and longitude lines. Of course, the figure Columbus was using for the circumference was incorrect, but once the earth's true circumference was determined by Jean Fernel in 1525, and its precise shape (flattened at the poles) worked out during the seventeenth and eighteenth centuries, the actual location of anything on earth could be plotted by simply determining its latitude and longitude. The scheme was elegantly simple—implementing it was the problem.

Determining where you are in the north–south direction (latitude) is relatively simple. Latitude could be determined simply by measuring the altitudes of either the sun at noon or (in the northern hemisphere) the constant pole star. At the equinoxes, the sun at noon is directly above the equator, and the farther north or south you go the lower in the sky it becomes. Already in the Middle Ages astronomical tables showed the height of the noonday sun throughout the year at different latitudes. Sailors could use a sighting instrument such as a quadrant or a cross staff to determine the height of the sun at noon and then, by referring to the tables, be able to fix their latitude to within half a degree.[4] Already in the early fifteenth century, sailors on Prince Henry's ships moving down the west coast of Africa had begun to mark their latitudes, and by the early sixteenth century maps showed quite precise latitude markings for many places along the African coast.

Longitude was another matter entirely, and it was longitude that needed to be determined once Columbus and the other explorers began to travel out into the oceans in the east–west direction. The earth's axis of rotation is the line running between the poles, so that they are the universal reference points for measuring latitude.[5] There is nothing like this for longitude. Not only does the east–west direction not culminate in any poles, it doesn't culminate at all. An arbitrary starting point (0 degrees of longitude) has to be chosen. It wasn't until 1884 that an international congress recommended that the whole world call an imaginary line running through the observatory at Greenwich, England, the prime meridian, or 0 degrees of longitude. Until that time, ships used a variety of prime meridians as their starting points for figuring out how far east or west they had traveled.

But *how* did they figure it out? The only way Columbus could do it was by knowing the latitude from which he set out, the direction in which he was going, and his speed. His navigational tools were extremely rudimentary—incredibly, he had only a magnetic compass for finding his direction, and the educated guesswork known as dead reckoning for determining his speed. Even the primitive logline method for determining speed would not be invented until the sixteenth century. Columbus reckoned his speed simply by watching bubbles and debris as they floated by his ship.[6] Although Magellan used celestial navigation three decades later, it was not yet developed in Columbus' time.

Given these primitive tools for finding longitude, it is no wonder that maps were so approximate and that, once found, places could often not be returned to. Above all, accidents continually happened because sailors miscalculated their proximity to shores, and when storms threw them off their courses they had no idea where they were. The Spanish even went so far as to have a rule of never sailing their ships at night for fear of running aground on something unknown. So prevalent and of such great impact on life were accidents at sea that their depiction became a whole subgenre of Dutch art during its Golden Age in the seventeenth century. Not only were lives lost and valuable cargo destroyed in accidents, but sailors also died of scurvy when destinations took too long to reach because of incorrectly reckoned longitudes.

18. *Ships in Distress on a Rocky Coast*, by Jacob Adriaensz. Bellevois. 1664.

The need to find a way to determine longitude with precision became even more pressing as the age of discovery turned rapidly into an age of far-flung international commerce. Within a century of the great journeys of exploration, the Dutch East and West India Companies were founded, the Spanish and Portuguese had colonies in South America, the English had their North American settlements, the French theirs in Canada and the West Indies, and the Dutch and Portuguese had trading posts in India and in the Spice Islands of Celebes and Malacca. Trade no longer went by slow-moving caravans across the land, but by ships sailing on the open oceans.

The first reward for determining the longitude at sea was offered by Philip II of Spain in 1567, and other rewards by Philip III, Louis XIV of France, and the Dutch States General followed. The problem was also an impetus for the founding of the Royal Observatory at Greenwich in 1675. After four ships of the Royal Navy carrying 2,000 men capsized off the Scilly Isles in 1707 because of a miscalculated longitude, members of the Navy as well as merchants sent an urgent petition to Parliament begging for the government's help in solving the longitude problem. In 1714, Queen Anne finally offered a huge reward of £20,000 (equivalent to several million dollars in today's money) to anyone who could find a way to calculate the longitude to within half a degree, i.e., thirty geographical miles at that latitude. The test would be a trip to Jamaica, where the longitude on land had already been determined by astronomical means. Even with that kind of reward, it took over fifty years for anyone to win it, and nearly *two hundred years* passed from the time of Philip II's reward offer in 1567 until a solution was finally found. Galileo tried, but failed, to get funding from both Philip III and the Dutch to develop his scheme to use Jupiter's moons. Newton spoke before Parliament about the huge difficulties involved in determining the longitude.

The expression "finding the longitude" became another way of saying "squaring the circle." In his image "Bedlam Hospital, London," Hogarth depicts many stereotypical lunatics, and one not so stereotypical. On the wall behind them is a drawing of the earth girded by latitude and longitude lines, and beside it an only slightly crazed-looking inmate making calculations to try to find the longitude. No matter that his efforts seem rational, to Hogarth the attempt itself was enough to signify his madness. In the history of human invention, there has been nothing else quite like the search for the longitude at sea.

It wasn't much easier on land, but it also wasn't a matter of life and death. And on land at least some distances between landmarks had become known, and the solid ground offered a steady base for a telescope or a pendulum clock. On the open sea there was nothing but endless water and sky to be seen from the swaying deck of a ship, where no telescope or pendulum could be relied upon.

Even the Greeks of the third century B.C. realized that longitude, unlike latitude, could be understood and calculated as a function of *time*.[7] If it takes twenty-four hours for the earth to turn upon itself (i.e., turn through 360 degrees), then every hour it rotates through fifteen degrees of longitude. More precisely, each degree of longitude amounts to four minutes of time. If you knew the time at the place designated 0 degrees longitude, and you knew the local time at your present position, then you could simply compare the two times and know your longitude. You wouldn't need to know your speed, you wouldn't need to worry that a storm had thrown you off your course. Without knowing anything else, you could simply check your two timepieces and know your longitude absolutely.

The telescope and pendulum clock mentioned earlier indicate the two forms this timepiece would take—one was celestial and the other mechanical. Both had been suggested by the midsixteenth century, and they continued to vie with one another until the mechanical clock finally won out in the late eighteenth century. The two celestial chronometers were the movements and eclipses of Jupiter's four moons (one of the first discoveries Galileo made after his invention of the telescope in 1609), and the movements of our own fast-moving moon against the background of fixed stars. Tables compiled over decades told the time of a particular movement or eclipse of one of these moons at what was designated as 0 degrees longitude. Navigators could determine the local time of the same event, compare it to the time at 0 degrees longitude, and translate the difference between the two times into the spatial terms of longitude.

After years of effort, both of these celestial chronometers were made to work. Galileo's idea of the Jovian eclipse system was used on land—in fact, the method was successfully used by the Cassini family to make a quite accurate map of France, and it was also how the longitude of Jamaica was established.[8] The lunar-distance method was used at sea. The drawback of both systems was that they required highly-trained mathematicians to do laborious calculations to correct for atmospheric refraction, the earth's motion, and several other variables. To make matters worse, all of this work would be rendered inaccurate if even a tiny error were made in the initial reading of the sky from which the calculations were done. Needless to say, such a tiny error was particularly likely on the deck of a rocking ship. Nevertheless, it was quite reasonable to look to the heavens for the answer to the longitude problem, particularly since the latitude could so easily be found there, and many people refused to believe that the best solution to the longitude problem was not to be found in the heavens.

The other approach dispensed with the heavens altogether and used a mechanical instead of a celestial clock. Instead of comparing the times of a particular celestial event at two different places, the strategy here was to keep *constant* track of the time at 0 degrees longitude so that it could be compared with local time. Finding the local time on the ship was easy. At noontime, when the sailors took a reading of the sun at its highest point in order to determine their latitude, they would also, obviously, know that their local time was noon. They could then simply take out the clock they had brought along with them, which was set to the time at 0 degrees longitude, compare the difference between the two times, and for every four minutes of difference know that they were 1 degree east or west of the prime meridian. A fast, simple, straightforward job which didn't require making mathematicians out of sailors.

But where was the clock? *That* was the problem. The clock didn't exist which could keep such precise time at sea, and there seemed to be insurmountable odds against it ever existing. Remember that at the equator each degree of longitude is sixty-nine miles, so that if a clock were off by as much as one minute after weeks and weeks of sailing, the ship's position would be inaccurate by over seventeen miles. Queen Anne's requirement that the longitude be accurate to within half a degree meant that the clock could be off by no more than two minutes after a trip to Jamaica (about six weeks) on the rolling sea, perhaps through violent storms.

The story of the creation of the clock that could find the longitude at sea requires a book of its own (and it has them). Here we can only offer the bare bones of what is the most flesh and blood of tales. Take one man named John Harrison (1693–1776), a carpenter and self-taught clockmaker from a tiny Yorkshire village, and add a lifetime of unremitting effort to make a clock which would solve the longitude problem, and you have the necessary ingredients for winning the prize. He did it by slowly whittling away at all those factors that caused timekeepers to be inaccurate.

John Harrison spent seven years building his first marine chronometer, known as H.1, and the finished product stood over three feet high and weighed seventy-one pounds. Nearly every source of friction had been eliminated, particularly in the new escapement: the famous, ungainly-looking but frictionless grasshopper escapement. The seemingly insurmountable problem of clocks running fast in winter and slow in summer because of the effects of temperature change on their metal pendulums or balance springs had also been eliminated. Harrison compensated for the temperature change by alternating brass and steel in the mechanism so that the two metals' different contraction and expansion rates canceled each other out. He also removed all unnecessary moving parts and found ways to get rid of or compensate for the influence of irregularities in the wheel train. H.1 was tested on a voyage to

Lisbon in 1736, but there is no record of how well it did. H.2, which took four years to build, and H.3, which took *seventeen*, were both about as big and ungainly in appearance as H.1. Neither of these clocks got sea trials.

Then came the historic H.4. Completed in 1759, it was a mere watch, with a diameter of 5.2 inches. In the more than three decades he had steadfastly devoted to making his four marine chronometers, Harrison had filed off so many of the sources of a clock's imprecision that when H.4 was tested on a trip to Jamaica and back during the winter of 1761–1762, it lost only five seconds on the trip out, which translated into a longitude error of only one and a quarter nautical miles! It was again tested on the return trip, which had day after day of terrible storms, but when the ship arrived back in England after a trip lasting a total of five months, its longitude reading still erred by less than the agreed-upon thirty miles.

19. H.4. The instrument that solved the problem of finding the longitude at sea was a 5.2-inch diameter chronometer which looked like a giant silver watch.

Thus, for the entire trip, H.4 had lost less than two minutes of time and was off by less than half a degree of longitude. John Harrison had earned the prize, even though it would take the personal intervention of George III for him to get all of the money in hand.

On his seventy thousand mile second voyage begun in 1772, Captain James Cook would use a copy of H.4 to help him both to find his way and to map the islands and continents of the Pacific. Today, the time given to the sailors of the eighteenth and nineteenth centuries by H.4 and its progeny is carried by radio signals, but the principle of comparing the time on the ship with the time at some other known place in order to determine longitude remains the same. The need to have precise time information in order to know *where* they are is equally great for the pilots of our airplanes and spacecraft. It will always be.

Time is bound to space and space to time, not only on our turning, spherical planet but for navigational purposes even beyond our planet. And knowledge of both is bound to an ingenious man-made machine called a clock, which was brought to a pitch-peak of perfection (for its day) by a strange, tenacious, self-taught man from a Yorkshire village. An event recalled by John Noble Wilford bears retelling. When Neil Armstrong, the first person to be carried to the moon by a modern navigational system, was being feted at 10 Downing Street, he proposed a toast to the man who started it all. His toast was *not* to an explorer, *not* to a statesman, and *not* to a scientist, but to a clockmaker named John Harrison.[9]

Chapter XI

Of Time and the Railroad

[The trains] go and come with such regularity and precision . . . that the farmers set their clocks by them, and thus one well conducted institution regulates a whole country.
Henry David Thoreau
Walden, 1854

Except that both stories involve travel on our spherical, turning earth, the story of time on land could not be more different from that at sea. There are no rewards in this story, and none of the single-minded effort to solve a life-and-death problem. On land, longitude had been successfully worked out using surveying techniques and the Jovian moon system. The various towns and cities already knew precisely what time it was where they were—ironically, *that* was the problem. On land, the difficulty presented by travel was in fact very nearly the reverse of that at sea. A town is not a moving ship, so on land the need was not to use time to pinpoint precisely where you were, but rather to understand that pinpoint accuracy can stand in the way when travel begins to demand greater communication between distant places.

Our story begins in 1784, when within a year's period of time the mail coach system was instituted in England. Let's first look for a moment to the way things were before 1784. Ever since mechanical clocks began to chime from church towers, Europe had been a jumble of different local times. Not only did some cities begin their day at noon and others at midnight, but each set its clocks not to a broad man-made zone of time as we do, but to the true sun time of its particular spot on the earth's surface. If you traveled to another city, you found out what time it was when you got there. When the pendulum clock arrived in the late seventeenth century, it had no effect on this hodge-podge; it just made each local time more accurate.

This was fine, since it was a world in which the fastest speed depended on the horse. Until the mid eighteenth century, travel on land was no faster than it had been in Julius Caesar's time, and many roads were so bad that wheeled traffic nearly ceased during the winter months, leaving people stranded in their towns and villages.[1] The only vehicles doing any traveling worthy of the name were ships at sea. During the eighteenth century, roads began to be tarred, and turnpike systems were built in both England and France. (Our toll roads today are still called turnpikes, after the spiked revolving frame that served as a barrier until the toll was paid.)

Without these tarred roads, the metamorphosis in travel (and hence in time) couldn't have gotten off the ground, but its real origin occurred in 1784, when a mail-coach system was put in place throughout England by John Palmer, MP of Bath. These coaches carried mail and also had room for passengers. What was really revolutionary about them was that they kept to a strict timetable. The drivers carried watches and the horses often died from being driven too hard in order to stay rigidly on schedule. Soon other coach services came into being, all with precise arrival and departure times. Their fierce competition was based primarily on the reliability and speed of their schedules.

Even though England is a small country aligned in a north–south direction, local time at places west of London can be up to twenty minutes behind it, and to the east seven minutes ahead. Until the mail-coaches came along, these different local times had presented no problem at all. Although the coachmen managed to set their watches ahead or behind to make them conform to the various local times, a problem which would plague travelers for the next hundred years had reared its ugly head.

It would take the arrival of the railroads in the second quarter of the nineteenth century to reveal just how ugly it was. With their greater speed, it was impossible for trains to match their time to all the various local times, and they tended to keep the time of the city from which they originally departed. The many railroad companies which sprang up within just a couple of decades had enough trouble keeping their timetables in synch with one another so that passengers could make connections; they could do nothing to prevent new passengers from missing trains because their clocks were set to local time. The problem was so annoying that clockmakers began to make watches with two dials, one for local and one for railway time.[2] In the United States, none other than Henry Ford designed such a watch.[3]

The railroads solved the problem among themselves by gradually agreeing to all keep London time, as Greenwich Mean Time (GMT) was called. By 1848, nearly all of England's railway companies were on GMT, which made them conform to one another but solidified the wall that had arisen between railway time and local time. In France, the railroads all kept Rouen time, which differed from that of Paris by five minutes, so that Paris railway stations kept their outside clocks five minutes ahead of those inside.[4] A similar bifurcation occurred on the rest of the continent and in the United States, and its resolution in tiny England was blissfully easy compared to the problems in a huge country like the United States.

Although there was some grousing about the provinces being lorded over by London, in very short order the English did the only sensible thing by taking the next step: they gave up their "real," local times altogether and set their town clocks to railway time. By 1855 nearly all public clocks in the British Isles were set to GMT,[5] which effectively put all parts of Britain into one time zone, although such a concept had probably not yet arisen in anyone's mind.

How were these distant cities able to know precisely what the Greenwich Mean Time was, and why was it such a paragon of accuracy? To answer the second question first: GMT was checked by astronomical observation of certain "clock stars" everyday at the Greenwich Observatory, and the Observatory's electrical Standard Clock was corrected to the new reading obtained from the stars. (By this time it was realized that a much more precise reading of our earth's rotation time could be obtained by recording the transits of certain stars at night than by trying to find the sun's zenith at noon.) The electrical telegraph had been invented in 1839, and by 1851 telegraph lines had been laid alongside the major railroad tracks. In 1852, a plan masterminded by the Astronomer Royal George Airy (1801–1892) was put into effect: time signals from the Observatory's clock were transmitted along the telegraph lines (in the form of electrical impulses) to electrical clocks and time-balls throughout England.

Railways, government and post offices, and of course telegraph offices received the time signals, and for a fee, private subscribers (usually jewelry stores and clockmakers) could also be hooked up to receive GMT. Think back five hundred years to when our forebears were profoundly ensnarled in the problem of the selling of time known as usury. It is mind-boggling to try to imagine what they would have made of this new wrinkle on the selling of time. The time signal was a product that was very profitably sold not only in England but on the continent and in the States as well, and it seems to have caused only a ripple of protest. But there were other ripples, *big* ripples, caused by these new manipulations of time, and the best place to study them is in that giant country where railway lines were being laid down across what had so recently been a new Eden, a new God-given Paradise.

In the United States as in England, the railroads were the primary motivation for the change to a more standardized time, but because of its size, the process was more complicated here. Total American railroad mileage increased ten-fold between 1840 and 1860, and at least in the East, trains were everywhere. As in England, each railroad tended to keep the time of the city in which the line originated—there were about eighty different timetables in use at mid-century. Needless to say, making connections between trains was chancy in the extreme. Consider the typical problem of a traveler from Portland, Maine:

"... on reaching Buffalo, NY, [he] would find four different kinds of 'time': the New York Central railroad clock might indicate 12:00 (New York time), the Lake Shore and Michigan Southern [railroad] clocks in the same room 11:25 (Columbus time), the Buffalo city clocks 11:40, and his own watch 12:15 (Portland time)."[6]

Late in the 1840s, monthly timetables coordinating the various lines began to be issued, but it was always an uphill battle for the traveller.

In order to further alleviate the confusion, regional time zones began to be put in place. By the early 1850s, all New

England railroads ran on the same time, kept accurate by daily telegraphic time signals from the Harvard College Observatory.[7] In the next two decades, other pockets of standardized time were created around New York, Albany, Cincinnati, Philadelphia, and Chicago, all of them controlled by time signals telegraphed from observatories.

A story about the Golden Spike, the last spike driven in the transcontinental railroad, reveals that the regional time zones were only a stop-gap measure for dealing with time in a world growing larger and more interconnected by the minute. The place was Promontory, Utah, and the year 1869. The railroad's backers had the idea of telegraphically transmitting the final blow of the sledgehammer across the nation in order to create the first synchronized moment in the country's history. It was a great idea, but the chaos of local times at which the event was reported to have occurred drowned out the intended sense of unity. We had grasped the need to bind the country together in terms of time as well as transportation, but weren't yet equipped to do it.[8]

The stumbling block was more intellectual than technical. It would require a change as dramatic as the shift from temporary hours to the clock's equal hours during the Middle Ages, or the acceptance of usury, but unlike them it would not be gradual. A closer analogy would be to Julius Caesar's calendrical revolution, but this one would be even more abrupt—the transformation of American time occurred at precisely 12 noon on November 18, 1883. Unlike Caesar's new calendar, standard time wasn't decreed from on high. It didn't even have the backing of the federal government.

Here, in brief, is what happened. By the 1870s, astronomers were coming up with plans to standardize time, but they didn't have the influence to put them into effect. The vast majority of people didn't travel enough to see the virtue of the idea, and even the railroads, focused as they were on profits, were content to let timetables help people navigate the crazy quilt created by the various railway times crossed with the local time of each station they passed through. Nevertheless, during the 1870s several national committees were formed to study ideas for standardizing time. The first plans concerned only the railroads—it seemed to be unthinkable to touch local time. Some people suggested putting all railways on a single time that would extend across the whole nation, but the plan which drew the greatest interest was that put forth by Charles F. Dowd (1825–1904) in 1869. Although Dowd originally aligned his four time zones to a Washington, D.C. prime meridian, he subsequently used the Greenwich meridian as 0 degrees, and made his four zones cover one hour each (fifteen degrees) using seventy-five degrees for the eastern zone, ninety degrees for the midwest, 105 degrees for mountain time and 120 degrees for the far west. Even Dowd wouldn't go so far as to suggest doing away with local time. Instead, he published timetables showing the time difference between local time and railway zone time for over 8,000 stations.[9]

Astronomers and even some railwaymen were more far-sighted; they conceived of giving up local time altogether and making railway time the *only* time. This was such a radical notion that it took a decade of discussions and lobbying of railway managers, city governments, and time signal merchants for a consensus to be reached among those who would have to be involved in implementing the change.

To help explain how it all came together, we need to look at an instrument most of us regard as just an oddity of New Year's Eve. Today we are literally surrounded by extraordinarily accurate timepieces—our quartz watches are more regular than the earth's rotation, and our telephone and radio time signals are connected to even more accurate atomic clocks. It wasn't like that back in 1883. On New Year's Eve, millions of people across the nation tune in to Times Square to watch a giant ball drop at the stroke of midnight. So far from being a modern oddity, this ball is actually a relic from the past. As early as 1833, such balls served as public time signals—to ships in harbors and rivers and also to people in towns. In 1861, there were time-balls not only in Greenwich, London, and Washington, D.C., but also in Liverpool, Edinburgh, Glasgow, Madras, Calcutta, Sydney, and Quebec.[10] Each of them was connected to an observatory so that they could deliver the most accurate time possible. By 1883, many major American cities had a time-ball, and people would stop for a moment at noon and correct their watches by it. It was the primary time standard to which the person in the city street could turn with confidence.

Previously, most time-balls had shown local time. If they could be made to reflect railway time instead, local time would instantly be abolished, at least for city people. That was part of what the lobbying of city governments, time signal merchants like Western Union, and even observatories was all about. The observatories had to agree to telegraph the new time, and city governments had to be willing to run their schools, courts, and other functions according to it. Western Union had to agree to sell the new time instead of the local time, which had always been its stock in trade.

Newspapers took stands for (and occasionally against) standardized time, and served as forums for a vitriolic debate on the issue. A fierce attachment to local time was found in surprising places—not only lay people but also many astronomers and even railwaymen clung to it with tenacity. The mastermind behind the ultimately successful effort to instate standard time was William F. Allen, a career railwayman devoted to its cause. Eventually, his lobbying efforts paid off: the observatories agreed to send the new time signal at precisely 12 noon on November 18, 1883. It would be disseminated to post offices, government offices, and time-balls, and the railway lines also issued precise instructions for changing their schedules.

It happened as simply as that. In a moment it was over, and a revolutionary new map of time blanketed the United States. Because it didn't involve the enactment of any law, there were of course pockets of dissent, but within a year eighty-five percent of all towns of more than ten thousand inhabitants were on standard time.[11] It wouldn't be nationally legalized until 1918, and then only as a by-product of the legalization of daylight savings time.

LOUISVILLE & NASHVILLE RAILROAD CO.

OFFICE OF THE GENERAL SUPERINTENDENT OF TRANSPORTATION

CIRCULAR No. 8.   *Louisville, November* 9. 1883.

Important Notice--Change of Standard Time.

TO TAKE EFFECT ON SUNDAY, NOVEMBER 18, AT 10 O'CLOCK A. M.   **8897**

On Sunday, November the 18th, at 10 o'clock a. m., the standard time of all Divisions of this Road will be changed from the present standard, Louisville time, to the **new standard ninetieth meridian or central time**, which will be eighteen minutes slower than the present standard time.

The system to be adopted in changing **Regulators, Clocks,** and **Watches** will be as follows:

On **Saturday, November** the 17th, at the usual hour for sending time—namely, **10 o'clock a. m**—all Clocks and the **Watches** of all **Employes** must be set to the exact present standard time. On **Sunday, November** the 18th, all Telegraph Offices must be open, and all Operators on hand for duty not later than 9 o'clock a. m., and remain on duty until relieved by the Chief Dispatcher of the Division.

All Work Engines and Crews, Road Masters, Supervisors, Section Foremen, and all other Employes who are required to have the correct standard time, must report at their nearest Telegraph Station not later than 9:30 a. m., on Sunday, November the 18th, and remain at the Telegraph Office until the new standard time is received, and their Watches are set to the correct new standard time.

On **Sunday, November** the 18th, the present standard time will be sent over the wires as usual, at 10 o'clock a. m., and as far as possible Dispatchers must have regular trains at Telegraph Stations at that hour.

At Precisely 10 o'clock a. m., by the present standard time, all **Trains** and **Engines,** including **Switch Engines,** must come to a **STAND STILL FOR EIGHTEEN MINUTES,** wherever they may be, until 10:18 a. m. by the present standard time, and at **precisely 10:18 a. m.**, by the present standard time, all **Clocks,** and the **Watches** of all employes must be turned back from 10:18 to exactly 10 o'clock, which will be the new standard time, and the new standard time will then be given from Louisville over all Divisions of the road.

20. Announcement of the time change for employees of the Louisville & Nashville Railroad. In this case, trains had to stand still for eighteen minutes in order to make the transition to zone time.

Here is how a writer in *Harper's Weekly* described it:

On Saturday, the 17th of November, when the sun reached the meridian of the eastern border of Maine, clocks began their jangle for the hour of twelve, and this was kept up in a drift across the continent for four hours, like incoherent cowbells in a wild wood.

But on Monday, the 19th—supposing all to have changed to the new system on the 18th—no clock struck for this hour till the sun reached the seventy-fifth meridian. Then all the clocks on the continent struck together, those in the Eastern Section striking twelve, those in the Central striking eleven, those in the Mountains striking ten, and those in the Pacific striking nine.[12]

And of course the minute hands were all identical across the nation, only the hour changed. Before there had been as many positions for the minute hands as there were places under the sun. The new system was orderly, rational, and practical. Why, then, were so many people so up in arms about it?

At every step along the way toward more standardized time, there had been problems of wounded regional and national pride. When the regional time zones had been implemented, small towns had resented having their own place in the sun (literally) usurped by Boston or New York time, just as English towns had grated when their "real" time was subsumed under London's. On a national level, many people were deadset against giving up a Washington, D.C. prime meridian for the one at Greenwich. And now that its local time was to be obliterated by a zone time initiating from Britain, a Louisville, Kentucky, resident asked: "Now what is Greenwich to us? A dingy London suburb."[13]

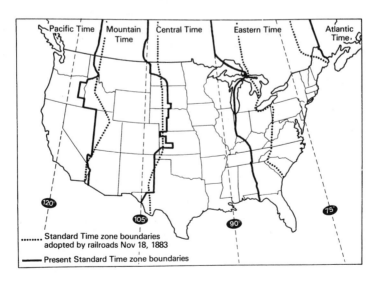

21. Standard time zones in the United States.

More interesting were the concerns that transcended such conventional problems of local pride, concerns that were remarkably similar to those surrounding the usury question five hundred years earlier. (These same concerns would be heard again in the debate over daylight saving time during the first World War.) Although there was also some hostility to the railroads themselves, in general the transformation in travel brought by the railroads was welcomed, while the transformation in *time* that it carried in its wake was profoundly troubling to many people. Travel by railway didn't tamper with any elemental "given" in the way that railway time (now standard time) did. Once again, time was seen as a primordial God-given phenomenon which ought not to be manipulated for any mundane purpose, certainly not for mere convenience in transportation and commerce. In his superb book on American time, Michael O'Malley has documented these feelings as they were expressed in newspapers of the day. A writer to the *Pittsburgh Dispatch* had this to say: "God Almighty fixed the time for this section just as much as he did for Philadelphia or New York."[14] Standard time was seen as false, a fabrication created by railwaymen who had lost all sense of the priorities. The feeling was that one might travel by railway time, but surely one could not be born, or married, give birth, or die according to something so synthetic. Some things ought to be left as nature brought them to us. Someone writing to the *Cincinnati Commercial Gazette* expressed it like this:

"The proposition that we should put ourselves out of the way nearly half an hour from the facts so as to harmonize with an imaginery line drawn through Pittsburgh is simply preposterous . . . let the people of Cincinnati stick to the truth as it is written by the sun, moon and stars."[15]

There is a problem here that has eluded the writer to the *Cincinnati Commercial Gazette*. There may ultimately be a "truth as it is written by the sun, moon and stars," but we are always only in the process of approaching it, and the truth at any point in time is half what we *believe*. Our citizen of Cincinnati may know that the earth revolves around the sun, but he approaches the problem of time from a Ptolemaic perspective. For him, the "truth as it is written by the sun, moon and stars" means the heavens revolving not even about the earth but about *Cincinnati*. He has forgotten that his clock already deviates from that truth by showing mean time rather than true sun time, and that it is only its synthetic twenty-four hours which allows him to fume about being put "half an hour from the facts." Most importantly, his Cincinnati-centered outlook ignores the fact that Cincinnati is a dot on a rotating planet revolving about the sun.

The practical problems created by the railroads were the push that forced standard time into existence, but the idea originated with astronomers. At every step of the way, railroad time was tied to observatories, which by definition scanned the heavens for "the truth as it is written by the sun, moon and stars." The astronomers had far more precise knowledge about that truth than any farmer's almanac or any Cincinnati resident checking the noonday sun. They looked at time in relation to the whole rotating earth instead of to just one point upon its surface. And because their view took in so much more, it had to be painted in the broader strokes of standard time. "Nature," "truth," and "the facts" are served at least as well by standard time as by local; they are just defined a little differently. Local time is just as man-made, just as much a cultural product, as standard time, and it is every bit as far from the elusive thing itself.

Chapter XII

The Worldwide Web of Time

At my dying hour, and over my long life,
A clock strikes somewhere at the city's edge.

Rabindranath Tagore

In 1883, only England, Sweden, the United States, and Canada had zone times based on the Greenwich Meridian. The rest of the world continued not only to be a disconnected mishmash of local and railroad times, but it was also still awash in prime meridians. Although the Greenwich meridian was now used to calculate longitude by about two-thirds of the ships at sea, the remaining third used ten different prime meridians.[1] And on newly published land maps during the early 1880s, Greenwich had no prominence at all. Each country still tended to use its own capital as prime meridian.[2]

In our day of instant global communications of every sort, it is hard to believe that this was the situation just a little over a century ago. But is it harder than to grasp that just a century before that the English mail coach was bringing the first presentiment of the complex connection between time and travel on land?

When a few far-sighted people like the Canadian railway engineer Sandford Fleming (1827–1915) began to come up with ideas about wrapping one unified system of time zones all the way around the world, they were met with the same incredulity that had earlier greeted the idea of doing away with local time. People simply had no concept of what the world was going to become in the next hundred years. One American geographer had this to say: "A capital plan for use during the millennium. Too perfect for the present state of humanity. See no more reason for considering Europe in the matter than for considering the inhabitants of the planet Mars."[3] Nevertheless, as was the case with American zone time, the far-sighted few plowed on, and eventually succeeded in foisting their ideas of temporal order upon an uninterested world.

Before worldwide zone time could even be thought about, all countries would first have to agree to use the same prime meridian. Every couple of years since the early 1870s, international congresses had been convened to discuss this question.

The Great Pyramid of Egypt, Jerusalem, and the Bering Straits had all been suggested as possible places for the prime meridian to run through, but it soon became clear that the 0 degree meridian would need to pass through a top quality observatory. That fact, and its predominant use at sea, gave Greenwich leverage.

At the International Meridian Conference in Washington, D.C., delegates from twenty-five countries put the matter to a vote on October 13, 1884. Twenty-two voted to make Greenwich the universal prime meridian, San Domingo voted against, and France and Brazil abstained. (France, always hung up on its rivalrous relationship with England, suggested that it might be won over to Greenwich if England would just accept its own pride and joy of measurement, the metric system.) This vote was only a recommendation to the governments of the participating countries, but it set in motion the changes that would bring us to the temporal world we take for granted today.

The most important consequence of the meeting (the adoption of Greenwich-based zone time around the world) was not even one of the resolutions on the agenda. Nevertheless, once Greenwich was accepted as the prime meridian for all the world, and with the highly successful system put in place in the United States during the previous year as an example, other countries began to follow suit. It was a gradual process that occurred country-by-country for the remainder of the century and on into our own. In many cases, it was first adopted by the railroads and only replaced local time later on. France couldn't bring itself to accept the Greenwich meridian until 1911, and did it then only by describing it as Paris Mean Time retarded by nine minutes, twenty-one seconds (in other words, *precisely* the Greenwich meridian). Liberia finally got around to adopting zone time in 1972—up until then its time had steadfastly remained forty-four minutes and thirty seconds behind that of everywhere else in its zone.[4]

However revolutionary, the idea of worldwide zone time was really quite straightforward: the 360 degrees of longitude which circle the earth were divided into twenty-four zones, each fifteen degrees wide. Like the system already functioning

1848	Great Britain (legal in 1880)
1879	Sweden
1883	Canada, United States (legal in 1918)
1884	Serbia
1888	Japan
1892	Belgium, Holland, South Africa
1893	Italy, Germany, Austria-Hungary (railways)
1894	Bulgaria, Denmark, Norway, Switzerland, Romania, Turkey (railways)
1895	Australia, New Zealand, Natal
1896	Formosa
1899	Puerto Rico, Philippines
1900	Egypt, Alaska
1901	Spain
1902	Mozambique, Rhodesia
1904	China coast, Korea, Manchuria, North Borneo
1905	Chile
1906	India (except Calcutta), Ceylon, Seychelles
1908	Faroe Islands, Iceland
1911	France, Algeria, Tunis and many French overseas possessions, British West Indies
1912	Portugal and overseas possessions, Samoa, Hawaii, Midway and Guam, Timor, Jamaica, Bahamas
1913	British Honduras, Dahomey
1914	Albania, Brazil, Colombia
1916	Greece, Ireland, Poland, Turkey (legal)
1917	Iraq, Palestine
1918	Guatemala, Panama, Gambia, Gold Coast
1919	Latvia, Nigeria
1920	Argentina, Uruguay, Burma, Siam
1921	Finland, Estonia, Costa Rica
1922	Mexico
1924	Java, USSR
1925	Cuba
1930	Bermuda
1931	Paraguay
1932	Barbados, Bolivia, Dutch East Indies
1934	Nicaragua, East Niger
~1936	Labrador
~1937	Cayman Islands, Curacao, Ecuador, Newfoundland
~1939	Persia
~1948	Aden, Bahrein, British Somaliland, Calcutta, Dutch Guiana, Kenya, Federated Malay States, Oman, Uganda, Zanzibar
~1953	Raratonga, South Georgia
~1954	Cook Islands
~1959	Maldive Island Republic
~1962	Saudi Arabia
~1972	Liberia

22. Dates of Adoption of Zone Times Based on the Greenwich Meridian

in the States, the time within each fifteen-degree zone was the same, and precisely one hour ahead of the neighboring zone to its west, and one hour behind the zone to its east. Throughout all of the twenty-four zones, the minute hands would remain the same. As more and more countries joined on, some deviations from the precise zones were made so that the same time could be kept within a nation, state, or group of islands in an otherwise empty Pacific, but the principle remained intact.

Sadly, the first World War did more to solidify world zone time than any of the efforts of scientists or politicians. Even before the war, instantaneous telegraphic communication was already outpacing the organization of time. Messages would arrive and there would be no way of knowing when they had been sent. Had they gone out before or after other information had been received by the sender? Without a uniform time system, there was no way to know. This resulted in real problems for diplomatic efforts prior to the war. When the war finally broke out, the need for coordinated time was all too apparent, and worldwide time thus received its greatest push from modern warfare's requirement of an organized temporal framework.[5]

But it would have happened even without the war. As the result of Marconi's discovery of wireless telegraphy in 1899, already in 1905 the first radio time signal began to be transmitted from Washington, D.C., to help ships find their longitudes, and it was soon followed by others in Europe. (One is almost thankful that John Harrison never had an inkling that a century and a half later his shipboard clock to tell the time at a known, distant meridian would be rendered essentially obsolete by a radio time signal.) The twentieth century revolution in communication had already begun, and it would require rational and predictable time around the world. What the railroads did to bring about the rationalization of time within individual countries during the nineteenth century, wireless communication would do for the whole *world* in the twentieth.

At first glance, our web of time zones seems utterly different from the discovery of how to find the longitude, and yet they were the two great steps in making our planet a single temporal entity. The time zones seem to be an imposed, fabricated structure compared to the technique for finding longitude—a decree rather than a discovery. A more accurate way to put it would be to say that both were steps toward the solution of a problem: how to understand time in the context of our whole turning earth rather than just for an isolated point upon it. No earlier culture ever determined longitude or invented time zones because no earlier culture had discovered the extent of our whole rotating globe and had also invented ways to move rapidly about on it. No other culture ever had any *need* to find a way to wrap time around the world.

Our time zones may be a man-made construct, but they are at the same time a response to the discovery that a map could be made of our planet's time just as maps had been made of its physical terrain, and that a temporal map was just as necessary for our movement about the earth. It may look like nothing more than an abstract grid, but it is still a map, and without it the various times of our turning planet would be an unnavigable Babel.

The invention of the clock began a gradual, centuries-long transition from a perception of time as something rooted in nature to something which originates in the clock itself. The creation of our time zones was a giant step in this transition. The earth's 360 degrees (the same 360 degrees that divide the clock face) were now separated into the twenty-four equal zones by which our clock time partitions our days. The turning earth had become a clock face and the time zones its numerals. Nature's earth clock had been transformed into a giant man-made clock, so that even the traveler never needs to look at the sky.

There is a paradox here. On the one hand, our clocks have caused us to forget the signs of nature in our counting of time. Yet from another perspective, to the same extent that they seem to have abstracted time from nature, clocks have entrenched it more deeply. When we change our watches without looking at the sky as we travel from time zone to time zone, we are actually expressing a far greater knowledge of time as it relates to our turning planet than were our forebears watching the sun above their heads.

Chapter XIII

A Watch on Every Wrist

*He looked repeatedly at his watch,
and when he had done so for the fifth time. . . .*
Henry James
The Ambassadors, 1903

We must now return to the machine itself. Look down at your wrist and you will almost surely see a watch. It won't be a mechanical watch; instead, it will be powered by a battery and regulated by the vibrations of a quartz crystal. This would not have been the case in 1884. You wouldn't have found a watch on your wrist at all—watches were still pocket watches (or sometimes in the form of brooches or pendants for women) and they had just begun to be something affordable by nearly everyone rather than being status symbols for the wealthy. They were all still mechanical—the far more accurate quartz crystal mechanism hadn't yet been heard of, much less the digital readout which often substitutes for the dial face today. The exponential increases both in the sheer quantity of timepieces and in their accuracy constitute the double revolution in time during the past century and a half.

In order for our concept of time as something to be found in clocks rather than in nature to become pervasive, it was necessary for man-made timepieces to thoroughly infiltrate our lives. This had been slowly happening all along since its invention, but during the nineteenth century it began a precipitous escalation. Let us focus on watches, since they keep clock time everpresent and inescapable in people's lives in a more pressing way than do clocks on the wall. The worldwide output of watches increased nearly ten-fold during the first three quarters of the nineteenth century: from 350,000–400,000 pieces a year to 2.5 million pieces,[1] and it has continued ever since. In 1978, 265 million watches were made,[2] and today output is over half a billion per year. About 300,000 watches are sold every day in the United States alone,[3] meaning that in the last decade of the twentieth century *daily* sales in the United States are three quarters the *annual* sales for the whole *world* just two centuries earlier.

Although watches have become fashion statements today, and cheap enough so that many people have more than one, the escalation was still primarily due to *need*—or, rather, to need as it brought about transformations in the means and philosophy of manufacture. Until the last half of the nineteenth century, the watch had always been a piece of jewelry—a work of art as much as a utilitarian machine. When half a billion are being churned out every year, that is obviously no longer the case. Even by the end of the last century, the watch had been reimagined along democratic lines, in terms of method of production as well as of the finished product. Let's briefly trace how and why this little machine took over the world.

During the eighteenth century, the British had dominated the market for clocks and watches, both in manufacture and consumption. Not only was Britain far more urbanized than any of the European countries, but its coach services, which preceded those on the continent, also gave it a greater time consciousness. As factories began to replace the putting-out system of labor in the late eighteenth century, more and more workers started and ended their work day at a particular hour of the clock.[4]

More than anything else, it was the rise of the railroads that caused clock time to begin to really circumscribe people's actions as well as their consciousness—not only in England, but in other countries as well. People began to want a watch not as a toy or piece of jewelry or status symbol, but in order to connect themselves with the proliferating time schedules in the outer world. The demand for clocks and watches swelled, and a changing of the guard occurred in what was already a highly competitive international market.

It was the Swiss who came to dominate the clock- and watch-making industry during the nineteenth century (and on through the twentieth until the new quartz crystal technology pulled the rug from under their feet). But their domination was only in manufacture, not in consumption. They were far too tiny a country for that. They had no choice but to export their products. This necessity forced them out into the international market and made them more finely tuned to its opportunities and challenges than was a country like England, which had a large home market. They also pulled ahead of the

British in their techniques of mass production, as well as in their pragmatical grasp that hand work should be just enough to fulfill the requirements of accuracy and appearance. The old British craftsmen couldn't give up their habits or their bias for the hand-made, and kept cutting by hand parts that a machine could cut better, and polishing and filing parts that didn't need it, thereby slowing down production and raising prices.[5] These were only a few of the complex factors that allowed the tiny country of Switzerland to dominate the world watch market for a century and a half.

The only hint of a challenge to the Swiss came from the United States. It began in 1850, when Aaron Dennison founded the Waltham Watch Company, which was followed by many other American firms during the coming decades. Although the American output was only a drop in the bucket compared to that of Switzerland, these companies introduced new mass production techniques which would ultimately transform watch production. Again necessity played a role: just as Switzerland's lack of a home market had caused it to develop export savvy, so the absence of a low-cost, skilled pool of labor in the States forced it to find ways to make machines do what the labor pool was doing in Europe. Where the Swiss used machines to cut or press watch parts and then hand-finished them, the Americans automated the finishing step as well. As the need for skilled human hands became less and less, the possibility of truly standardized, truly interchangeable parts came into view for the first time.

This was difficult enough to do in clocks, so how did the Americans manage it in something as tiny as a watch, where the necessary fit between parts often required measurement to the ten-thousandth of an inch or less? To the Swiss way of thinking, the only way to fit together a balance wheel and hairspring (or jewel and its pivot) was by the intervention of a skilled craftsman who could adjust the delicate parts so that they would fit together like hand and glove. The Americans tackled the problem from a completely different angle: their machines stamped out or cut huge numbers of parts as close to identical as possible. These were then sorted according to their hairline differences and placed in different containers. A relatively unskilled worker could then take from the right containers so that the tiny deviations from the standard accommodated one another—a slightly larger pivot for a slightly larger jewel, and so on.[6] In 1876, after they began to lose market share in the States, the Swiss finally decided to check the accuracy of these machine-made watches. They were stunned to discover that the "inferior" American products kept far better time than their own hand-finished timepieces. Not only were they more accurate, but they could be made and sold at far less cost.

By this time, the need which had been triggered by the railroads had grown a life of its own, and many other factors fed into it to create a huge market for affordable watches. During the Civil War years, one facet of that market was the American soldier. Like all wars, this one had particular need for time synchronization, and it helped Dennison's company to prosper by providing a huge market for its inexpensive Ellery model, which even a soldier could afford. By 1865, this cheap watch accounted for nearly half the company's sales.[7] Waltham, as well as the other companies, suddenly became aware that it wasn't only soldiers who were buying their inexpensive watches. Lots and lots of other people wanted affordable watches, and the American companies responded to this vast market by producing cheaper and cheaper watches throughout the remainder of the century, culminating in the Ingersoll Company's famous "dollar watch" in 1898.

From the 1880s on, the best place to buy these inexpensive watches was not in the local jewelry store, but in a mail-order catalogue. Sears, Roebuck entered the world as the R.W. Sears Watch Company in 1886, sending out into rural America page after page of watches from which the farmer could choose.[8] Farmers and people in small towns had the least practical need for watches and had been among the most vociferous in their antagonism to standard time (and would be again to daylight saving time), but they sure wanted clock time of some sort. Their desire for watches is a telling index of the growing primacy of man-made time even in rural life.

In the city, a spur to watch ownership came with the first electric lighting in 1882, when Thomas Alva Edison (1847–1931) hooked up his new light bulbs to an underground cable system carrying direct current electrical power. At 3:00 P.M. on September 4, 1882, four hundred electric lights came on in offices on Spruce, Wall, Nassau, and Pearl streets in lower Manhattan. Soon after, across the nation and in France, Germany, Britain, Italy, and Holland, franchises for Edison's power system were bringing man-made light to the world.[9] As the arrival of electrical lighting dissociated our work/rest cycle from its immemorial basis in nature's day and night, so it also irrevocably ruptured the connection of time-telling with that natural alternation. When night was nearly as bright as day, more than ever one needed a watch to find one's way in the temporal world of the city night.

Yet another change gave the urban working person a motivation for owning a watch nearly as powerful as the soldier's. We have spoken about the rise of the factory system of labor without asking how people managed to get to work on time or how the hours they worked were determined. With the earlier putting out system of cottage industry, workers had been paid by how many parts they produced, but when they moved into the factory, their pay began to be determined by how many hours they spent there. As early as 1844, the Englishman Charles Babbage (1792–1871) had invented a mechanical time clock, and variations and improvements upon the theme had continued ever since. By the turn of the century, time card stamping systems similar to those we still use today had been patented. Sears was not the only great American company to get its start in the time-keeping business. What we know as IBM began as the International Time Recording Company, which during the first decades of our century had a virtual monopoly on the manufacture of time clocks in the United States.[10] With time cards commonplace, life had finally become sufficiently regimented by the clock for a watch to be part of the survival equipment of all but the lowliest worker.

But it wasn't yet a *wrist* watch! The first wrist watches were decorative ornaments made for women, and men conse-

23. Advertisement for a time card system by the International Time Recording Company (later renamed IBM), 1911.

quently shunned them as being too effeminate. Again war had its effect—this time it was the first World War. A watch that has to be fished out of a pocket in order to be read is a lot less convenient than one strapped to the wrist, particularly when you're trying to fire a machine gun or charge over a hill at a particular moment. Realizing this, governments made wrist watches part of the standard equipment issued to their soldiers. These watches looked a little different from ours because they were covered with a metal grid rather than glass—unbreakable glass hadn't yet been invented. (The growing dependence of the world on clock time is further indicated by the fact that Civil War soldiers had to buy their own watches, while half a century later they were issued to World War I soldiers as part of their equipment.) After the war, men suddenly felt that wrist watches were acceptable for civilian wear, a trend which clearly has continued for both sexes until the present day.

The rest is history. The forces of advertising and mass production and need have compounded to bring us to our present half billion per year output, and there is no end in sight.

The range of watch types has grown like an evolutionary tree. At first there were just two great branches: the fine, status symbol watch to be handed down to one's offspring, and the inexpensive, utilitarian type to be thrown away. During the 1950s and 1960s, the Timex was the mid century's version of the dollar watch. Its case couldn't even be opened for repair; it was cheap enough that when it stopped you just threw it away and bought another.[11] Today's version is the quartz crystal watch easily assembled by cheap labor in Hong Kong or elsewhere in Asia and sold by a multitude of companies. But already in the 1930s, advertisers were trying to steer people away from the one-watch habit by encouraging them to think of watches as appropriate for different activities. The first sports watches—for golf!—date from the 1930s.[12]

Today we still have the watch as status symbol and as pure utility, but in between endless branches have sprouted on an ever-growing tree. There are sports watches to time anything. There are watches for divers, watches for pilots, watches for sailors. There are watches with every conceivable add-on: calendars, perpetual calendars, moonphases, alarms, time around the world. There are digital read-outs as well as the traditional analog dials. Most interesting of all is the watch's casework, which has developed into a real democratic art form. It can express anything from technological wizardry to elegance to outright whackiness. The tree has so many branches now that it is hard to see through them to its old dualistic structure.

Just because watches have become bearers of much nonchronometric meaning isn't to say that they're not here primarily because we *need* them. We need our watches now more than ever. Airline and bus schedules have been added to those of the railways, and electromagnetic waves bring us radio and television broadcasts for which we want to be on time. Everything in our world runs by the clock—from the stock market to parking meters. We cannot conceive of life without it.

Consider our efforts to gain an understanding of life in ancient Greece or Rome. We have what remains of their literature and art, and we have a conception of their forms of religion and government. We try to feel our way into their world, and we persuade ourselves that we can to some small extent, until we are stopped short by the absence of clocks and watches. What can the ordinary person's day have been like? What did time *feel* like to them? We can imagine their emotions and philosophy of life more easily than we can conjure the effect of that wrist watch-less looseness of time.

Chapter XIV

New Vibrations

*In our own age, even a second can be
regarded as a great clumsy piece of time.*

J.B. Priestley

During the twentieth century, the stupendous escalation in the number of clocks and watches was more than matched by their increase in accuracy due to the advent of quartz crystal technology. And that was just the beginning. The great saga of the invention and development of the mechanical clock, a story that took nearly seven hundred years to unfold, is over. It is one of the great tales of our culture, but it is now part of history. Although there will always be a market for "mechanicals," there is no way around the fact that their buyers will be people wealthy enough to indulge an antiquarian taste.

Except in terms of style, our wrist watches today (at least those with hands and dial) look essentially the same on the outside as did those of the early part of the century. But inside—no one could guess it was the same machine. In the digital watch, there are no moving parts at all. There is nothing to see but a tiny battery, a tiny capsule containing a quartz crystal, and a tiny bit of electronic circuitry. And in the dial watch, out of all the mechanical clock's exquisitely attuned vibrating and turning parts, only the wheel train leading to the dial remains. It is no use pretending that these are clocks in the sense that they have always been understood, and yet the *principle* of the very first mechanical clock is retained in this strange, foreign article. If their *content* is totally different, their *form* remains the same: the battery generating electricity has replaced the weight or mainspring as the source of power, the quartz crystal has taken the place of the verge-and-foliot, balance wheel, or pendulum as the time divider, and an integrated circuit counts the crystal's vibrations, reduces their thousands of beats per second into our language of time, and then transmits this language to the clock face.

How did this change come about? No one went out searching for this revolution in time-keeping. It came in by the back door, so to speak, but once discovered, its virtues were all too apparent. Although no one knew it at the time,

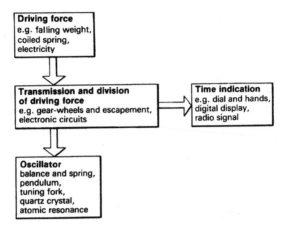

24. Essential elements of all clocks.

the demise of the mechanical clock was already augured as early as 1880, just about the time mail order catalogues were beginning to bring the watch to the American hinterlands. Pierre Curie (1859–1906), working in his laboratory in Paris, discovered piezoelectricity: when pressure is applied to certain crystals (including quartz crystals), they can be made to vibrate at a very constant frequency. (Piezo derives from the Greek for "to press.") It was soon found that an alternating electrical current served even better than pressure to make the crystal alternately expand and contract at an extraordinarily steady rate. Beginning in the 1920s, this strange phenomenon was used in radio broadcasting. Quartz crystals kept broadcast waves on a set frequency. The vibrating crystal generates a weak electrical current of the same frequency as its vibrations, and this can be used to keep the frequency of a broadcast wave constant.

If the vibrations of a quartz crystal were stable enough to keep radio waves on a fixed frequency, couldn't they be used to regulate a clock? This question was asked by W.A. Marrison of Bell Laboratories, and he answered it in 1928, when he built the first clock regulated by the vibrations of a quartz crystal. Its accuracy—within one or two thousandths of a second per day—was far greater than anything a pendulum clock could ever achieve.

Why? For precisely the same reason the pendulum itself had proved so superior to the verge-and-foliot: its vibrations are far more stable and accurate. A pendulum has a natural periodicity within the earth's gravitational field, yet many factors work to upset its constancy. Recall that a metal pendulum rod expands in summer, causing it to oscillate more slowly and thus lose time, while in winter it contracts and its faster oscillations make the clock run fast. A pendulum is also slowed by friction with the surrounding air, and this becomes greater when the atmospheric pressure increases. The invention of gridiron pendulums made of two metals whose differing responses to temperature cancel each other out, and the placement of the pendulum in a vacuum chamber, helped to alleviate these two problems. However, a clock's pendulum is housed within a complex mechanical apparatus, subjecting it to further causes of friction.

Ever since the pendulum clock was invented in 1657, ingenious efforts had been made to isolate the pendulum from the other parts of the clock so that it could tick away solely in accordance with the natural law built into it. Ironically, the greatest breakthrough in freeing the pendulum from extraneous perturbations came in the early 1920s, just a few years before the first quartz crystal clock was built. Bringing to fruition an idea that had been around since the turn of the century, W.H. Shortt built what became known as the Shortt free-pendulum clock. Shortt took the ultimate step in his attack on interference to the pendulum's free movement—he simply put the pendulum in a separate clock. Shortt's clock was in fact *two* clocks, a "master" and a "slave," connected by electricity. The slave clock did the job of generating the power for the pendulum and counting its oscillations, leaving the solitary pendulum to swing in a partial vacuum with only the tiniest push to keep it going. With all the potentially interfering work done by the slave clock, the master's freely oscillating pendulum attained an accuracy to ten seconds per year.

Shortt clocks were installed at Greenwich in 1924. By 1939 they had already been superseded by quartz crystal clocks—not only at the Observatory but in terms of timekeeping accuracy. Although Shortt clocks were accurate to within ten seconds per year, this still put their accuracy a hundred-fold behind that of quartz crystal clocks. Despite Shortt's brilliant work, the most stringently isolated pendulum in the world couldn't compete with something that hardly needs to be isolated at all. The problems that plagued the pendulum were simply thrown out the window by the quartz crystal. It is only slightly responsive to changes in temperature and virtually inert to differences in atmospheric pressure. No extraneous mechanical functions perturb the regularity of its vibrations or cause them to be mistranslated enroute to the clock face. The steady alternating current from the battery causes the crystal to vibrate, and the integrated circuitry translates its vibrations into the reading we see on the dial or digital readout. That's all there is to it. What it lacks in human artifice it more than makes up for in scientific accuracy.

The second reason the quartz crystal is so much more accurate than the pendulum has to do with the frequency of its vibrations. We have already seen that the faster things vibrate, the smaller the periods into which they divide time, and thus the greater their precision as timekeepers. The unit of time we call the second existed only in theory until the 39.1-inch Royal pendulum was invented to measure it. Until then, nothing could beat off so small a bit of time. The quartz crystal can vibrate thousands (even millions) of times a second; a mechanical clock can't measure bits of time this small, but today's electronic devices easily can.

To early humankind, the smallest unit of time was the day—the earth-sun clock can't measure anything smaller. Then for millennia the smallest unit was the hour—first the temporary hour of sundials and later the equal hour of the mechanical clock. By 1500, public clocks were beginning to strike on the quarter-hour, but the minute could not be accurately counted until the pendulum clock, and the second had to await the invention of the 39.1-inch pendulum in 1670.

With the quartz crystal clock's ability to measure thousandths of a second, time counting underwent another reversal. Just as mechanical clocks superseded sundials after they revealed irregularities in the sun's daily passage, so now the quartz clock was able to measure irregularities in the earth's rotation rate. It had long been known that the earth is slowing down due to friction between ocean tides and shallow sea floors caused by the gravitational pull of the moon. (If proof were needed that the earth is a clock, perhaps its problem with friction will suffice!) In addition to this long-term slowing, two other sources of rotational change exist: one irregular and unpredictable, and the other seasonal. The irregular fluctuations occur because the molten core and solid mantle of the earth rotate at different rates, and the seasonal changes are the result of freezing and melting of the polar ice caps, which affects the atmospheric pressure and causes the earth clock to run fast in the fall and slow in the spring and early summer.[1] There are also seasonal changes in the direction and velocity of the prevailing winds, and their friction with the earth's surface creates small fluctuations in the earth's rate of rotation.

Once quartz clocks had revealed these irregularities, it became clear that the earth's daily rotation could no longer serve as the basis of timekeeping. It didn't make sense to define the second as 1/86,400th (24 × 60 × 60) of a mean solar day when it obviously wasn't. Sometimes it was longer and sometimes shorter. But it was equally clear that if it couldn't be time's source, the earth would forever remain the central fact to which any new, improved time would have to accommodate itself. What we knew as GMT is now called Coordinated Universal Time (UTC). It is administered from Paris instead of Greenwich, and it times the daily crossings of incredibly distant quasars rather than mere stars, but its raison d'etre is still—as it has always been—to determine the earth's time relative to the heavens, irregular or not. As we will see, the perfect second would have to come to terms with this reality, and not the other way around.

Before we look at the clock that would replace our earth clock in the twentieth century, we must reflect a moment on why it was needed. What may seem to be a crazed, obsessive search for a timekeeping Utopia was in fact a very necessary, practical effort. Almost without knowing it, we have segued into a new frame of reference in terms of time. Suddenly we are talking about *seconds* rather than hours and *vibration frequencies* of crystals instead of swings of pendulums. We have moved into the tiny end of the measurement scale just as we have exponentially expanded our ability to communicate and move across our planet and beyond. It is no coincidence that these two seemingly opposed movements have occurred together. Our electric power and our communication and navigation systems all require the coordination of minute frequencies, and this cannot be accomplished without absolutely precise timekeeping.

Think of our electrical power networks, each of which involves hundreds of generators. To get a sense of such a modern power system, imagine a hundred heavy pendulums strung together by delicate threads. If the pendulums swing perfectly in time, the threads will hold. If they begin to swing at even slightly different frequencies, the threads will break and the whole system descend into chaos.[2] That is what happened in the historic blackout along the east coast of the United States on November 9, 1965. In order to prevent such catastrophies from happening, the frequencies at power stations must be checked against a frequency standard many times more precise than even they need to be. Today that standard is an atomic clock.

In terms of broadcast waves, consider the following: An FM station is assigned the frequency of 100 megahertz (1 billion oscillations per second). If the station's second differs by only 1/1,000th from the true second, its broadcast oscillations will be off pitch by 100,000 hertz (cycles per second).[3] Our communication networks literally run on electromagnetic waves, and their smooth working requires a time standard reliable to the thousandth, millionth, or billionth of a second.

The effort to discover the longitude revealed the first dim intimation of the need for precise timing in travel through space. We have already told how the Apollo 12 crew set up reflectors on the moon so that laser beams from the earth could be bounced off them and reflected back. By multiplying the time required by the speed of light, the distance could be determined. Now, if the precision of the timing is within one part in a million, the distance can be determined within 415 yards, but if it is within one part in ten billion, the distance can be determined to within two *inches*![4] Navigation of our space flights and our telescopic searches into the distant universe also require this kind of precision timing.

This brings us to the shortcomings of even the quartz crystal clock. The definition of a second which ultimately replaced that of 1/86,400th of a day was not based on the quartz crystal's vibrations, which are extremely stable short-term, but begin to drift over long periods of time. This "aging" of quartz crystals has several causes: the slight changes in frequency brought on by temperature changes, impurities within the crystal, and the accumulated effects of the vibrations themselves.[5] This drift may amount to only milliseconds within a month, but even that is too great for many of today's precision instruments.

Although vibrating quartz crystals seem worlds away from swinging pendulums, they both function as the natural oscillator at the heart of their respective clocks. They are alike in another way as well: their frequencies are not inherent to the material they are made of, but depend on the *size* to which that material is cut. Just as the same metal pendulum will beat for different time periods if it is shortened or lengthened, so a quartz crystal's vibration frequency depends upon the size and shape to which it is cut. Crystals can vibrate a few thousand or many millions of times per second, and, in general, the smaller the crystal is cut, the higher will be its resonant frequency. Even with crystals we can't escape the eternal problem of all things mechanical: no two man-made things can be cut exactly alike. If its frequency depends on the width to which it is cut, then the quartz crystal is already this side of perfection as a clock, even without its aging problem.

Didn't there exist somewhere in the world a natural frequency that would never run down? And wasn't there somewhere a frequency that was a property of the material from which it resonated and not of the manner in which that material was cut? Incredibly, this doubly difficult criterion was met during the twentieth century. By 1910 Ernest Rutherford had proposed a model for the atom that made it a kind of miniature solar system, with electrons orbiting a nucleus. By 1913 Niels Bohr discovered a new law governing the behavior of matter and energy: quantum mechanics. In the larger world, friction causes everything to eventually run down. Even the planets are very gradually losing energy, slowing down, and spiraling toward the sun. However, within the miniature solar system-like world of the atom, another set of laws seems to rule. One would have imagined that the electrons, circling the nucleus like tiny planets, would also lose energy and spiral inward. Not only wasn't this happening, but something else entirely *was*: the electrons lost or gained energy not gradually, but in discrete lumps (or quanta) as they jumped from one definite orbit to another or even as they changed their magnetic alignment while spinning within the same orbit. This jumping from one energy level to another went on and on forever; there was no eventual running down. The perpetual motion machine, billed as a Utopian impossibility in the larger world, is in fact exactly what atoms are.

Not only did atoms not run down, but the frequencies with which they resonated were intrinsic to their nature and not the result of any cutting or shaping from the outside. The energy released or absorbed when an electron changes energy levels takes the form of radiation at a frequency unique to that energy level change in that particular atom. Thus, the atom appears to be the one "device" in the world that fulfills the two criteria for a perfect timekeeper.

Atoms of many different elements can play the role of pendulum in atomic clocks, but the atom ultimately selected for use in defining the duration of a second was an isotope of the cesium atom, cesium-133. At room temperature, cesium-133 is a silvery metal, and at *any* temperature all cesium-133 atoms have a nucleus containing 54 protons and 79 neutrons, with 54 electrons spinning in orbits (energy shells) around the nucleus.

There is one lone electron in the outermost energy shell. All cesium-133 atoms are *exactly* alike. If one were to be found with so much as one electron too many or too few, it would simply no longer be cesium-133. Only on nature's production line, and only at the atomic level, is this absolute identity possible, and it is just this which explains the superb timekeeping of atomic clocks. There is no opportunity for friction, distortion, or aging—in identical atoms, the frequency emitted or absorbed when an electron changes energy status is also identical. If this kind of precision could be harnessed to a clock, it would be the greatest timekeeper the world had ever seen.

Although a molecular clock which counted frequencies associated with the ammonia *molecule* (NH_3) had already been built at the U.S. National Bureau of Standards in 1949, the first *atomic* clock, using the cesium atom, was constructed by the English physicists L. Essen and J. Parry in 1955. We won't describe the rather complicated working of this clock here, but we must explain what it counted, because this has become the most important bit of time in the world. Simply put, the electron in the outer energy shell of the cesium atom spins on its axis, as does the nucleus, and in both cases a magnetic field is set up. When these two magnetic fields are aligned in the same direction, the two spins are cooperative and the atom is in a slightly higher energy state than when the two fields oppose

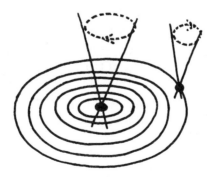

25. Our second is now determined by the difference in energy state of the cesium-133 atom when its nucleus and outer electron spin in the same, or opposite, direction.

each other. When the outer electron flips its magnetic direction relative to that of the nucleus, it emits or absorbs a tiny quantum of energy in the form of radiation with a frequency of 9,192,631,770 cycles per second.

However, these electrons can go without flipping for tens of thousands of years. The cesium clock is simply an ingenious device to force these outer electrons to flip and keep on flipping so that their radiation frequencies can be counted. In essence, it works like this: The cesium atoms are zapped with microwaves of very nearly 9,192,631,770 cycles per second. This causes their outer electrons to flip, and a feedback loop is set up. As the cesium electrons flip at their natural frequency, they in turn hold the microwaves to the same frequency. The microwaves are thus locked onto the cesium frequency, and the clock can count off perfect seconds of 9,192,631,770 cycles.[6]

When it was invented in 1657, the pendulum clock's gain or loss of only ten seconds a day made it a paragon of accuracy. Three hundred years later, the cesium clock is accurate to a few billionths of a second per day, making it nearly ten billion times more accurate than the pendulum.[7] In 1967, the 13th General Conference of Weights and Measures formally agreed upon a new definition of a second: "9,192,631,770 periods of the radiation corresponding to the transition between the two hyperfine levels of the ground state of the caesium-133 atom." Now, instead of the top-down approach of defining the second as 1/86,400th of a mean solar day, a new bottom-up definition had been put in place. The day had officially become 86,400 atomic seconds, and the atomic second the single fundamental unit of time.

The origin of timekeeping in the clock rather than in the turning earth was no longer merely perceptual; it was now literal. This momentous change involved something else as well: It meant that this new second—an atomic second governed by the laws of quantum mechanics—had replaced the gravity second determined by laws controlling such things as a pendulum's movement in the earth's gravitational field. Now timekeeping would have the new task of finding a way to fit this new, improved time to the awkward earth.

Here was the problem. The exquisite accuracy of atomic clocks is what stands behind the power companies' ability to keep the electricity to our homes and offices at precisely 60 cycles per second so that our lights and electric appliances can run. It is also the frequency standard without which our televisions, radios, computers, and all our communications systems could not work. When we call to get the correct time, the answer is hooked up to an atomic clock. Modern society is literally held together by electronic technology that routinely communicates by signals synchronized to billionths of a second. "The 'time' as most of us know it is simply inexpensive crumbs from the tables of the few rich 'gourmet' consumers of time and frequency information."[8] Our quartz crystal watches are good enough for individual use, but we are never far from the "true" time based on the atomic second. Now the fact of the matter is that this time system into which we are all hard-wired requires only the *uniformity* of its time standard; it doesn't care about the rotation of the earth. Such a macro-movement simply does not exist for it.

But it exists for *us*, and most particularly for navigators and astronomers. The rotating earth isn't about to go away just because it doesn't quite fit with the vibrations of cesium atoms. It was soon discovered that the time on cesium clocks differed from mean solar time by as much as a second a year. (Just as there aren't an even number of days in a year, so there aren't an even number of atomic seconds in a solar day.) This may seem trivial, but once again modern technology proves that it isn't. "In electronic navigation, a time error of a millionth of a second can produce a position error of about a quarter of a mile. Get your celestial timing wrong and spacecraft will sail past planets, missiles can fall in the wrong places, and jets can land short of the runway."[9] Thus, in 1972 it was decided to insert (or delete) a leap second whenever atomic time and UTC differed by more than nine-tenths of a second. This way, our frequency-based technological world can have the invariant second it requires, and our time can still stay in synch with the planet which brings us night and day and the changing seasons.

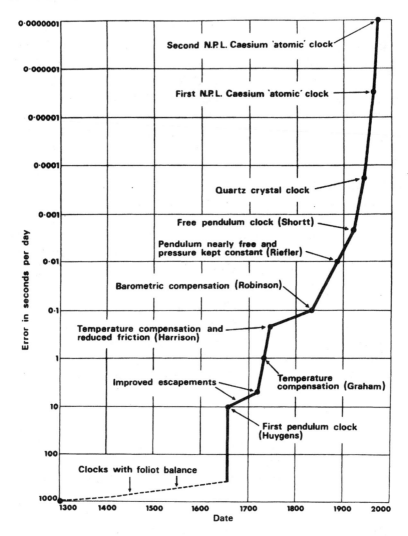

26. The increasing accuracy of clocks, from the first mechanical clock to today's atomic clock.

Throughout the previous pages, we have set up an opposition between the man-made time of the clock and the natural time told by the apparent movement of the sun and stars as the earth revolves upon its axis. Although there is certainly truth to this dichotomy, it is interesting that the clock's greatest improvements in accuracy have always occurred after a bit of nature became its counting device—first the pendulum, then the quartz crystal, and finally the cesium atom. Now, with our atomic clocks accommodated to the turning earth by leap seconds, it feels more as if nature writ small has become bound to nature writ large than that the man-made has accommodated itself to nature. There is now a cesium clock accurate to *one second in three million years*[10]—such a phenomenon doesn't result from man-made, but only from man harnessing nature.

Chapter XV

The Clock and Charles Darwin

But by and by they came to my watch, which I had
hidden away in the inmost pocket that I had . . .
They seemed concerned and uneasy as soon as they got hold of it.
Samuel Butler
Erewhon, 1872

Samuel Butler's imaginary tribe of Erewhonians, fearing that machines could evolve into conscious beings and rise up to destroy their human creators, ceased to use them and hid them away in museums. When Butler's hero leaves his sheep farm in New Zealand to wander over the mountains into Erewhon, he is arrested for the crime of carrying a pocket watch.

Despite the ever-increasing skill of our chess-playing computers, we clearly don't fear that the rapidity with which modern machines have evolved forbodes their eventual leap into consciousness. And yet, aside from that last step, the Erewhonians' use of a Darwinian model for their vision of mechanical evolution is not so far-fetched:

> Take the watch, for example; examine its beautiful structure; observe the intelligent play of the minute members which compose it; yet this little creature is but a development of the cumbrous clocks that preceded it; it is no deterioration from them. A day may come when clocks, which certainly at the present time are not diminishing in bulk, will be superseded owing to the universal use of watches, in which case they will become as extinct as ichthyosauri, while the watch, whose tendency has for some years been to decrease in size rather than the contrary, will remain the only existing type of an extinct race.[1]

Amazingly, the sundial is not yet extinct, but the water-clock is, and the temporary hours by which they charted the day for the ancient and medieval worlds certainly are. Mechanical clocks will surely one day be all but extinct, and the time may even come when something will replace the atomic clock.

But there is more to the Darwinian model than sequential extinctions. There is the retention in each successive species of vital characteristics inherited from its predecessors. When we strap on our quartz crystal wrist watches each morning, we are attaching to ourselves a scientific instrument which has encoded within it a heritage extending from deepest antiquity to the recent past.

It is the ancient world that we see on the clock face: the division of the day into twenty-four hours—originally twelve for day and twelve for the night—derives most probably from ancient Egypt. The division of the hour and second into sixty equal parts we owe to the sexigesimal mathematical system of the ancient Mesopotamians.

When we look inside our watches, we see a mechanism whose ancestor originated in Western Europe in the late thirteenth century. The creators of this first mechanical clock kept the ancient world's numerical divisions, but made all the hours equal to one another and, above all, based the measurement of time on something that oscillated at a regular interval rather than on something (like sunlight or water) that flowed.

The difference between the mechanism of our wrist watch and that of the earliest clocks we owe to the science and technology of the modern world. By the seventeenth century, the natural periodicity of the pendulum had given the clock a time divider governed by natural law. In our own century, the vibrations of quartz crystals and cesium atoms, although infinitely more precise, are likewise determined by the laws of nature. Electricity has replaced the gravity-powered crown wheel, and electronic circuitry supplants the wheel train, but the organizing principle remains unchanged.

Another recent mutation in our timekeeping concerns the way we think about time rather than the mechanism we use to measure it. For over a century, our watches have told us zone time rather than local time. Wherever we are, our watches let us know that we're not an isolated dot on the surface of the earth but a part of a whole turning planet. For all practical purposes, local time is now "extinct."

As we will see in Part II, for all the kinds of daily time that have been brought to extinction by the science and technology of our world, they have also given us new "clocks" to help us locate ourselves not within one diurnal rotation of our planet, but within the earth's whole lifetime. By 1872, just thirteen years after the publication of Darwin's *Origin of Species*, Samuel Butler could use its concepts and assume understanding from his readers. As we will learn, Darwin plays a role in the story of time that is factual rather than metaphorical, but this new chapter in time counting will *not* follow the Darwinian model of slow evolution. In this case, science will bring us abruptly into a new, fact-based vision of time which carries no inheritance from the myths or beliefs which preceded it. Its only inheritance will be the idea, largely a legacy of the clock, that time, which brings about change, can itself only be measured by counting a period that is constant and changeless.

Part Two

The Time of the Earth

Chapter XVI

Of Clocks and "Clocks"

*Any repetitive phenomenon whatever, the recurrences
of which can be counted, is a measure of time.*

G.M. Clemence

In the preceding chapters, a story that spans at least 4,000 years has been told, and yet the tale itself is devoted only to the present moment. Of all the clocks that humankind has invented, not one has been able to hold onto the time just passed. This is not the case with other forms of measurement. If you measure a board and find its length to be one yard, you can always take the board out and measure it again. You can weigh yourself one day and do it again the following week; check the height of your child this year, and again the next. You can't do that with time. The "evidence" simply evaporates. There seems to be no matrix to which the dimension of time "sticks."

With the atomic clock, the precise measurement of the present moment has gone about as far as it can go. Like a farmer using terraces on a hillside to squeeze every last bit of use out of his soil, we have domesticated time by cutting it into smaller and smaller pieces in order to make it as productive as possible. But in doing so, we have not escaped our confinement in the moment. We have made the present ever more fleeting with each smaller piece into which we have cut time.

Where could we go from here? What kind of an encore could there be after the atomic clock? The encore we got was so surprising and unexpected that no one could have imagined it. It has changed our idea of the world forever, and its echoes still reverberate in our ears. To say it more accurately, the music is still playing, for the same new knowledge of the atom that brought us to the pitch-peak of perfection embodied in the cesium atomic clock has carried us into a new and utterly different realm of time. The cesium clock does not represent a dead-end in time-counting; rather, the atom that gave us the cesium clock has also allowed our time-counting efforts to turn a corner, move away from a focus on the present and out into the vast reaches of time itself.

Until late in the nineteenth century, the clock that counts the time of day was the only phenomenon which went by the word "clock." Clocks took many forms, but they all did exactly the same thing: they counted the time within one revolution of the earth by dividing it up into twenty-four hours, each containing sixty minutes of sixty seconds. If you said "clock," that was the only thing you could be referring to. The clock on the wall or on the wrist is still anyone's first thought when he hears "clock," and yet seemingly every few weeks the newspaper tells us about a new discovery involving biological clocks, molecular clocks, genetic clocks, or radiometric dating (the counting of time by the constant decay rates of radioactive atoms). The greatest of all the new clocks, radiometric dating, somehow didn't get clock as part of its conventional name, but it is nevertheless the *sine qua non* of them all.

What happened to cause the efflorescence of these strange new clocks in the twentieth century? Although it isn't known who first applied the term clock to something other than an apparatus for telling the time of day, it couldn't have been done until the *concept* of clock had been pried loose from the machine itself. From its inception clear through to the end of the nineteenth century, the clock was one and the same as the machine. Clock was a *thing* as much as an *idea*. But finally the concept of clock became liberated from the one machine with which it was associated, and today we don't bat an eye when we read of these new clocks (and there will surely be many more of them in the future).

The new clocks represent discoveries of modern science, and they certainly don't depend on the traditional clock for their existence, and yet the concept upon which they are based had no other origin. Could today's radioactive clocks or biological and molecular clocks have come into existence without the prior millennia-long development of the clock to tell the time of day? Technically, the answer may be "yes," but in reality it is surely "no." In the first place, something as scientifically sophisticated and as precise as a radioactive clock just wouldn't have been developed by a culture which perceived its daily time in the crude terms of a sundial. More specifically, if the idea of using repetitive recurrences of identical changes to count the passage of time hadn't already existed, the time-counting potential of the newly discovered phenomena might never have been realized. Radioactivity, for example, might have been put to its other great uses without its possibilities as a clock being perceived.

In discussing the planetary basis of our day in the first chapter, we mentioned that the earth is 4.5 billion years old. How do we know? And it takes only a quick step into any library to learn the detailed timeline of our earth in millions and billions of years. Again, how can we possibly know? We accept these facts as if they were our birthright, and yet no one before this century had any notion of them. What has allowed us to have this precise knowledge of the earth's lifetime? The single most crucial answer is the radioactive clock.

In the following pages we will take a look at early ideas about the earth's time, and then we will see how the miraculous clock of radioactivity laid out before our eyes the vastness of our planet's history. Without this clock, we would still be mired in the guesswork and controversy that confounded students of our planet's history until late in the nineteenth century. These students—the great geologists, physicists, and biologists who discovered the "deep time" of our planet—still had no real idea of how much time they were talking about. That answer could only be found by something based on the same fundamental principle as the clock that gives us the time of day: time can best be measured by finding a source of regularly recurring identical changes and counting them.

For all their vast differences, this shared concept makes traditional clocks and radioactive clocks part of the same story. Their temporal domains may be at opposite poles from one another, and yet they both light up the time of their respective domains so that it can be seen. It is as if time only becomes visible, and hence real, by being measured.

Still, the clock counts a tiny wheel of time, an infinitesimal pinprick against the million- and billion-year-long half-lives of the radioactive elements which measure the earth's lifespan. Their differences don't end there. Take the issue of invention versus discovery: Although our clock incorporates much knowledge about the earth and the natural laws that govern it, it is nevertheless a wholly artificial, invented machine. The segments into which it divides time can be found nowhere in nature, and might easily have been otherwise. The clock was *not* discovered. Something very nearly the reverse is the case with the natural clocks which tell us the ages of things. We finally got around to discovering them early in the twentieth century, and all our inventiveness has gone into learning how to read them. Like the frequencies of the pendulum, quartz crystal, and cesium atom, the constant decay rates of radioactive atoms represent another case where nature herself provides a far more dependable time counter than anything humankind could ever devise. Here the natural recurrence is not a mere regulator put into a man-made instrument. Instead of being translated into artificial divisions of time, the natural recurrences of the radioactive clock stand on their own. They aren't just the heart of the clock, they *are* the clock.

Yet another difference lies in the disparate purposes of the two clocks. The traditional clock was conceived as a practical tool to help us make good use of time in our daily lives, and nearly all of its improvements have been devoted to that purpose. The natural, radioactive chronometers aren't practical at all—they are completely irrelevant to everyday life. Yet they have revolutionized our *knowledge* in the same way that the clock transformed our practical lives, and they have done it not in 4,000 years but in less than a century. And unlike our clocks, whose time demarcations occlude our vision of the thing itself, radioactive clocks not only don't let us forget the real thing, but they bring it to us with an awesome specificity. We say "It's 4:37 P.M." without a shred of awareness that this is a shorthand code for the position of the rotating earth relative to the sun; the clock has taken the place of the reality it represents. It's just the opposite when we say "The earliest ancestors of man don't appear to go back much more than 4.5 million years," or "The dinosaurs became extinct sixty-five million years ago," or "Few fossils of animals with shells have been dated much farther back than 550 million years ago," or "This 550 million years is only about thirteen percent of the whole lifetime of earth." So far from dimming or domesticating reality, the radioactive clocks—without which we could not know any of these facts—hold up before our eyes a timeframe too stupendous for our minds to grasp.

We said at the outset that clocks are perforce dedicated to the present moment because time doesn't seem to stick to things in the way that other dimensions do. Could there be a clock that disproves this rule? When we say that the earth is 4.5 billion years old we have already answered the question. The discovery of the radioactive elements has given us a unique kind of clock which is not forced to float upon the present, but instead can lift the traces of the deepest past out of rocks and bring them to us with a surreal precision for which we have no need other than to know and to marvel.

Our tale of the radioactive clocks will take a different shape from our earlier story. The saga of the traditional clock gave us a long, continuous journey through four millennia, with interest in the various timekeeping devices alive at every step of the way. There just isn't any comparable story for these new clocks. They enter their tale right at its end, solving an impasse in our knowledge which could *never* have been solved without them. They arrive by happenstance—even the greatest minds working on the problem of determining the true ages of events on earth could never have conceived that something like this would come to their rescue. So in order to tell this story, we will first need to take another detour—this time a long one—to look at the views of time and the earth that were buzzing in people's heads as they went about fine-tuning their sundials and waterclocks, or later as they rushed to work at the sound of the bells in the communal clock tower, or later still as they ran to catch a railroad train. This story isn't about technology, or about a culture's invention of a particular kind of daily time, but about the *real thing*—the real primordial phenomenon that we try to stay in contact with despite the coercion of our clocks and watches to keep our minds bent forever on the present moment.

Chapter XVII

The Circle and the (Short) Line

Unto the place from whence the rivers come,
thither they return again.
Ecclesiastes

The poor world is almost six thousand years old.
William Shakespeare
As You Like It

Every culture lives within its dream.
Lewis Mumford

A clock just *counts*—accurately, precisely, but nevertheless blindly. Before there can be any perceived need for a clock, there first has to be something that requires counting. In the case of the traditional clock, this was hardly a problem. The recurrent round of the day was obvious to even the dimmest early hominid, and it was entirely relevant to his life. On the matter of the age of the earth, no such "given" existed. Unlike the in-your-face phenomenon of the day, the earth's age had first to become something of sufficient interest even to be thought about.

That people have always thought about time doesn't mean they have always thought about the age of the earth. In fact, not until well after the astronomical revolution of Copernicus and Galileo did the idea slowly begin to arise that the earth was a distinct entity within the rest of the universe and that it might have formed at a different time than the universe itself. Until then, the whole phenomenon was seen as one entity, and whatever beliefs about it were held, were held for the whole. We are most familiar with this view from the first words of Genesis: "In the beginning God created the heaven and the earth." Throughout nearly all of the four thousand years that the clock was making its slow, yet relatively linear progress toward more precise control of the time of day, the time of earth, when it was thought about at all, was seen as the age of *everything* and was subject to a variety of conjectures, which bore no linear relationship to each other whatsoever.

It was not until the nineteenth century that enough of an empirically-based vision emerged—and enough disagreement about that vision—for people to begin to dream of some kind of a miraculous clock to set things straight. For the first time, there was something to be counted. Yet even this situation was very different from the phenomenon of the day. We didn't know *how much* time we were dealing with, and as it turned out, even the boldest conjectures were woefully off the mark. If a clock could someday be found, it wouldn't have the job of simply dividing up something already known, but would have to *discover* its domain even as it counted it.

"How old is the earth?" It seems such a simple, natural question. A child's question. Surely it must have been asked again and again from time immemorial. It wasn't. It really wasn't seriously asked until the mid eighteenth century (and wasn't answered until the mid twentieth). One can understand why no ancient Greek asked any questions about radioactivity, or evolution, or galaxies. They didn't know they existed. But they knew about the earth and time and age just as well as we do today. Why, then, were there no questions about the age of the earth—not from them, nor from the Egyptians or Babylonians or Romans? And not from any of the civilizations of the East or from anywhere else in the world that we know of? The explanation doesn't lie with their lack of means to answer the question, because the nineteenth century, so full of yearning to find an answer, was equally devoid of reliable means to attain it. And lack of a way to get solid, provable answers has never stopped the human race from speculating.

Consider the Greeks, those most inquiring of all ancient peoples. They had theories about the originating principle of the world (Thales believed it was water, Anaximenes thought it was air, and Heraclitus said it was fire), about the nature of the world (Parmenides saw its essence as one, unchanging reality, while Heraclitus conceived it as pluralistic and constantly changing), and about its physical make-up (Democritus and Leucippus believed it was all made of atoms and void). They were also intensely interested in time, and developed many diverse ideas about it—but they never asked about the age of the physical planet they lived on. And this despite the fact that for the most part they believed explanations for the big questions of life were to be found in physical mechanisms and not in myths. It was also in spite of their observation of the puzzling phenomenon of marine fossils located on mountain tops and their correct guess that at one time the strata in which the fossils were embedded had been under water. They made the observation but it didn't go much further than that.

It wasn't just that our forebears lacked enough of a sense of how nature "works" to even begin to answer the question, but rather that other assumptions and beliefs obstructed their vision so that they didn't think to ask. More than the means, they lacked the motive. There are surely equally simple, childlike questions lurking beyond the pale of our consciousness today about which future generations will wonder "Why didn't they ask that question?" We didn't because we didn't know enough to ask it or we thought that we already knew the answer or that the question was not worth asking.

Don't imagine for a moment that ancient civilizations didn't think about time. Time, like space, has always been one of the great questions, and every culture this side of the most rudimentary has devised an explanation for it. So let's go back to the first two thousand years of the development of the clock, and look at the thoughts on time itself that engaged the minds of our forebears as their sundials and waterclocks were slowly drawing them toward a man-made regimentation of their daily lives.

Forget for a moment the human inventions of the clock and of history, and look around at the natural signs you unconsciously use for ordering time. There is the ever-recurring day and the ever-recurring round of seasons in the ever-recurring year. Look up and you will see the repeating phases of the ever-returning moon. Beyond the moon, there is the seasonal pattern of the stars repeating itself every year and, with a little more effort, you could discern the recurring patterns of the planets. Nearly all the signs of nature, at least at first glance, tell us that the world repeats the same patterns over and over again.

Although there are signs to the contrary—an individual person is born, grows old and eventually dies, never to be reborn—the predominant natural signs reveal a world returning again and again to where it was before. It was from these signs that the earliest cultures derived their notion of time. Time, too, must be cyclical. It, too, must regenerate itself into an ever-renewing present.

We think of our own conception of time as linear, and yet, at the level of the day-to-day, we still live in a world of circular time. In fact, the clock is our own man-made cyclical time machine, not only keeping us constantly in the present, but bringing us, with each twenty-four hours, to the new time of a new day. With each dawn, the past is forever abolished. Lacking the idea of history, why shouldn't the ancient cultures have seen the broader reaches of time in the same way that we experience the time of day?

As Mircea Eliade has shown in his great book *The Myth of the Eternal Return*, the prehistorical cultures saw time itself as following the model of the recurring cycles of nature, and in one form or another, nearly all of the great ancient cultures continued this idea of time. The Babylonians, the Chaldeans, and at least the predominant pagan philosophies of the Roman world—the Stoics, Pythagoreans and Neoplatonists—all saw time as cyclical, and this idea was one of the thickest threads in the tapestry of time woven by the Greeks. The view wasn't confined to the West. The Chinese, Hindus, and Buddhists also conceived of time as recurrently renewing itself, as did all the Mesoamerican cultures in the Western hemisphere.

During two thousand years, and with so many different cultures involved, there were countless variations on the theme. The Babylonians and Buddhists had myths about the creation of their cyclical time, which would then continue forever. Plato also seems to have believed that time began with the creation of the world, while Aristotle saw time (and the world) as utterly eternal, without either beginning or end. Whatever its belief about the origin of time, each ancient culture had a unique idea about the period of time after which one cycle would end and a new cycle begin. These Great Years were based on planetary motions, calendrical periods, or a complex interweaving of both, and their durations ranged from a few decades to far longer spans of time. In addition to longer cycles, the Chinese had a sixty-year cycle based on the conjunction of Jupiter and Saturn in the same sign of the zodiac, and all of the Mesoamerican cultures had a fifty-two-year cycle, after which the days of their seasonal and sacred calendars would once again mesh in the same way. The Babylonians and many of the Greeks believed that time and the world would begin again when the sun, moon, and all of the planets again returned to positions they had held relative to one another at a previous time. The time cycles of the Indians are like wheels within wheels: A single cosmic cycle has 4,320,000 years, a thousand of these cycles constitutes a day in the life of Brahma, another thousand a night. Thus, a single day and night in the life of Brahma is longer than our own fifteen-billion-year-old universe. A hundred *years* of these billion-year-long days and nights makes Brahma's whole life, and after his death the greatest of the Indian Great Cycles will begin all over again.[1]

Some of these cultures believed that a world-wide conflagration or flood ended each Great Cycle, literally destroying the old world so that time and the world could be reborn as fresh as on the first day of every other cycle. The Indians and the Mesoamerican cultures believed in such a destruction, while the Greeks seem to have thought that their time would simply quietly begin anew once the planets made a complete cycle through the sky.

We tend to think of a conception of time as cyclical as something belonging to the distant past and utterly foreign to our own (correct) linear view of time. But this was never a tale composed in black and white. Many Greek philosophers may have thought of time as cyclical, but Herodotus, the original historian, was also a Greek. Other views ran like rivulets through the land of cyclical time, just as the reverse occurs today. We believe time to be linear in nature, and yet every New Year's Eve we celebrate the passing of old time and the birth of a new year. Our noisemakers are relics of a very ancient practice aimed at frightening away evil spirits from the clean and pure new time. We still have philosophers of cyclical time (like Friedrich Nietzsche), a cyclical stock market, and an oscillatory theory of the universe. As we will see, the cyclical view has never left us—it played a crucial role in early speculations about the earth's geological changes.

There is yet another story which reveals that the cyclical view does not belong exclusively to the distant past. Of all cultures that have *ever* existed, the Maya were the most obsessed with time. In addition to their fifty-two-year cycle, they also had an ongoing Long Count to tally the time within a Great Cycle, which amounted to approximately 5,125 of our years. Their current Great Cycle began on the same day as August 13, 3114 B.C. of the Gregorian calendar. Sadly, we are not the only ones anticipating a great turning point in time. Only twelve years after our new millennium begins, on December 23, A.D. 2012, their present Great Cycle will end,[2] the universe will be destroyed and born anew in the next Great Cycle. But who will know? December 23, A.D. 2012, will be an echo out of the long past, but only a handful of Maya adepts will be there to hear it. The Maya of the classical period, carefully inscribing Long Count dates on all their monuments, could never have believed that their time would die. They believed that time itself was *their* time. But it wasn't, anymore than it is ours.

Like all ideas, the cyclical vision of time is a tunnel at once leading directly on to other ideas while excluding those which lie to its sides. A question about the age of the earth, or any other question about the true extent of past time, lies to the side of the tunnel dug by the idea of cyclical time. It is simply nonsensical. If time is an endless series of repetitions, any sense of beginning is eclipsed, and without a beginning the question of absolute age is absurd. Even if the age of the earth could be determined within one cycle, what meaning would that have if the cycle ultimately ends and time begins all over again?

Not only did the Greeks sense the long spans of earthly time implied by marine fossils found on mountain tops, but they were also aware that over long time the ocean waves and rivers erode the land. To them, these were curiosities of the current cycle of an endlessly repeating cosmos, and that grand vision occluded from their minds the meaning we find in these elementary geological facts.[3]

Ultimately, we owe the meaning we see in these facts to the new, linear vision of time originating with the Jews and later disseminated throughout the Western world by Christianity. The word *ultimately* here is fully meant, because herein lies one of the strangest, most circuitous tales in all of history—it would be a long, long time before this debt could be perceived. In the beginning it was another story: "The grand impersonal sweep of the Greek . . . cycles of time was replaced, in Christian thought, by an abbreviated and anecdotal conception of the past, in which the affairs of men and God counted for more than the inhuman workings of water on stone."[4] As we will see, it would take nearly two thousand years for this inauspicious beginning to find its way back to a new vision of "the inhuman workings of water on stone," a vision which would finally begin to tell us the age of the earth.

Whether we are believers or not, those of us raised as Christians or Jews are all, even today, too close to the biblical account to quite grasp what a disaster it was for the empirical study of nature. We learn its stories in childhood, and the great tale of God's creation of the world in six days sinks into a place inside us too deep for the Greek explanations of natural causes to ever reach. Try as we may, our image of the biblical view in the context of the ideas of contemporaneous cultures will never quite lay flat. The picture always warps and buckles, and the biblical view pops out as if it were somehow larger than any of the others. It isn't larger, but for nearly two thousand years it won the day. What meaning could natural causes have beside Divine intervention? God was not just another originating principle to set beside the water or air or fire conjectured by the Ionian philosophers—God was the Creator.

The pagan cultures didn't ask the age of the earth because their cyclical view of time made such a question meaningless. The Jews and Christians didn't ask because they already knew how to get the answer: all the generations of Adam's descendants as told in the Bible had to be added up. Although Adam lived for 930 years, he was 130 when his son Seth was born. Seth begat Enos when he was 105, and Enos was 90 when he begat Cainen, and Cainen was 70 when he begat Mahalaleel, who was 65 when he begat Jared, who was 162 when he begat Enoch . . . . And on and on. Because of differences between the Greek (Septuagint) and Hebrew versions, this was not a straightforward task, and the job got a lot more difficult when the specific names ran out (at the reign of Solomon)—now events in the Kingdoms of Israel and Judah had to somehow be correlated to get a single chronology. The Bible couldn't be used at all for the four hundred years from Ezra and Nehemiah to the birth of Jesus, so that this period had to be counted according to dates in Chaldean and Persian records until these could eventually be correlated with the Roman dates which would give the time of the birth of Jesus.[5]

It was a difficult job, and its uncertainties account for the discrepancies among the figures obtained. The calculation was done many times over, and by some of the most brilliant people who ever lived. Theophilus of Antioch dated Creation to 5529 B.C., Julius Africanus calculated it to be 5500 B.C., the Venerable Bede got 3952 B.C., Martin Luther got 4000 B.C., Sir Walter Raleigh, 4032 B.C., J.J. Scaliger, 3950 B.C. and Isaac Newton, 3998 B.C. Because until recently the Irish Bishop James Ussher's date of October 23, 4004 B.C. was incorporated into the King James version of the Bible, it is by far the most famous. At least in the Christian world, this form of answer to the question of the earth's (and the world's) age stopped all other inquiries for nearly two millennia. For most of that time, it went totally unquestioned, and even after the first tentative questions began to appear, so great was its power that new discoveries ended up being bent and contorted to fit into its timeframe.

Because it is so far afield of what we know today, this answer to one of the great questions of life may seem childlike. For the sixteenth and seventeenth century calculators (Scaliger, Newton, Ussher and many others) this effort to determine the age of the world was just one aspect of a heroic attempt to bring together all of Western antiquity into a unified chronology. Using astronomy, mathematics, languages, coins and many other guides, these men began the enterprise which today allows us to go into any library and find information on any ancient culture we wish in temporal context with its surrounding world. By bringing the disparate dating schemes of past cultures under the umbrella of the B.C./A.D. system, they were creating the "clock" of universal history, whose unit is not the hour or the second, but the year. For this reason, and because the heretical notion of time as eternal was the only other conception then known, Bishop Ussher doesn't deserve the hindsighted ridicule which is often heaped upon him.

To return to the origins of this view of time: it was their exclusive concern with "the affairs of men and God" that brought the biblical writers through to a linear conception of time. Unlike the Greeks, they didn't set out with any particular interest in time per se; it simply followed of necessity from their human concerns: the great events portrayed in the Bible couldn't be belittled by having them repeated in future cosmic cycles. The Creation happened only *once*, Moses went up to Sinai to receive God's Commandments only *once*, and Jesus died on the Cross just *one time*. The whole meaning of the Bible would be destroyed if such awesome events were to be recycled. Out of the biblical writers' dedication to human stories and the personal bond between humankind and God rose a linear view of time. Time couldn't repeat; it had to go on and on.

It is nevertheless a great mistake to think that this view of time bears much resemblance to our own. Its similarity begins and ends with its linearity. In the first place, there was none of our own assumption of either historical development or natural causation in this early sense of linear time. "For both Christians and Jews, the history of the world was not the slow unfolding of a continuous development, but a sequence of unique events, each of which broke abruptly with all that had gone before."[6] They broke abruptly because they were the result of Divine intervention. The biblical deluge was caused by God's wrath; Abraham failed to kill his son Isaac because God commanded him not to, and not because he made a decision on his own; Christ died on the Cross because it was foreordained by Divine prophecy. It was a view of causation as foreign to us as it would have been to Aristotle.

Something else made this early concept of linear time almost as distant from our own view as is the idea of cosmic cycles. It was perceived as having an end as abrupt as its beginning, again resulting from Divine fiat. At the turn of the third century, the Christian writer Julius Africanus had built upon earlier periodization schemes to come up with a world composed of six ages, corresponding to the six days of Creation. These six ages made up a Great Week, which covered the period from the Creation to the end of the world as we know it. In the seventh age, Christ would return to reign on earth until the Last Judgment, after which the true believers would obtain "life everlasting" in eternity.

"For a thousand years in thy sight are but as yesterday when it is past . . ." Based on this line in Psalm 90, each of these ages came to be considered a thousand years in length, giving the earth a life expectancy of six thousand years. When Shakespeare's Rosalind mourns that "the poor world is almost six thousand years old," the audience in the Globe theatre understood it as a way of saying that the world was in its old age and that time was short for both it and its human cargo. (Our own view of time has changed utterly, and yet our upcoming millenial turning point has its origins here.)

Although it waxed and waned, speculation about when the world would end remained alive from the early years of Christianity clear through to the seventeenth century. Because of the differences among the conjectures about when Creation occurred, there was an equivalent variety of guesses about its end point. Hippolytus, Bishop of Rome, seems to have made the first calculation of A.D. 202, and the second was of A.D. 500[7] (recall that Julius Africanus dated Creation to B.C. 5500). These years came and went without the great event occurring, and it was not until A.D. 1000 that the next serious flurry of concern about the End arrived. This in turn was followed by Joachim of Fiore's calculation that the End would occur between 1200 and 1260.[8] Martin Luther, who had originally predicted A.D. 2000, believed the world had become so degenerated that it would not make it through the sixth millennium, but might perish as early as 1560.[9] And in the early seventeenth century, when John Donne tried to coax his coy mistress with the words "Had we but World enough, and Time," his meaning was the same as Shakespeare's.

Certainly not everyone was obsessed with the imminent end of the world. Among other things, the mechanical clock had been invented shortly after Joachim of Fiore's date for the world's demise, and Galileo's discovery of the pendulum came not long before John Donne's words were written. (If the moment of the end were to be recorded, the mechanism for doing it accurately was proceeding apace!) The situation recalls the controversy about the selling of time. In that case, hardheaded worldly types found ways to get around the Church's usury laws in order to forge ahead with their commercial ventures. Here, too, predictions of the world's demise didn't stop those same types from going ahead with their inventions, explorations, and discoveries. After all, the End had been predicted many times before and it hadn't happened yet. But this reasonable, life-loving approach was all they had to go on— they simply had no way to imagine an alternative to the world's death by an act of God analogous to the Creation with which it began.

Whether people obsessed about the end of the world or not, the Creation story with its start date of around 4000 B.C. was simply a "given" for everyone. Our spatial world had been exponentially enlarged: first by Columbus' discovery of the Western hemisphere and the voyages of exploration which followed, and then by Copernicus' displacement of the earth from the center of the universe and the transformation of heavenly bodies into other worlds wrought by Galileo's telescope. Nothing like this had happened with time. If we were no longer at the physical center of the universe, its time had still been made just for us—Creation predated Adam and Eve by just six days.[10] The Jews and Christians had invented the idea of linear time, but it appeared that it would take a Houdini to escape from the strictures imposed on it by the authority of God's Word.

The Greeks thought in terms of natural causes, but their time conceptions precluded any approach to determining the age of the earth. The Bible supplied the requisite linear time, but with the time barrier imposed by its Creation story (and Divine intervention as its modus operandi), it was equally incapable of approaching the problem. Before the question of the age of the earth could be answered (or even asked), a recombination between these two would have to take place, allowing natural causation to become aligned with the concept of linear time denuded of its biblical strictures. Only then could "the inhuman workings of water on stone" be perceived against a backdrop of linear, directional time, and the physical earth's stupendous antiquity come into view.

Chapter XVIII

Discovering the "Dark Abyss"

*These [fossils] are the greatest and most lasting
Monuments of Antiquity, which, in all probability,
will far antedate all the most ancient Monuments of
the world, even the very Pyramids . . .*
Robert Hookes (1635–1703)

*. . . we find no vestige of a beginning.—
no prospect of an end.*
James Hutton (1726–1797)

Every school child knows of the Copernican Revolution, which seemingly overnight brought us to an entirely new vision of space and our place in it. It didn't happen overnight, of course, and yet, between publication of Copernicus' treatise in 1543 and Galileo's views of other earth-like planets (with their own moon-like satellites) through his telescope in 1609, it seems to have been accomplished. At least for people aware of these discoveries, the boundary of the tidy, rigid Ptolemaic universe was irrevocably shattered, and space began the awesome process of opening out upon itself that continues to this day.

Not only do school children not have in their minds a comparable image of revolution in our conception of time, but neither do most adults. Perhaps time is somehow more abstract to us, so that the analogies created for it are more difficult to fathom. Certainly there is nothing in the discovery of time comparable to the clear and explicit expulsion of the earth from the center which inaugurated the enlargement of our spatial world. And there will be no markers for time quite so compelling as the stars and planets which return every night. The drama of the sky speaks to everyone, and it was surely the reason why astronomy became our first science. From the earliest beginnings until today (and forever) the planets and stars are the built-in coordinates to help people imagine whatever vision of space their culture gives them.

There is also man-made visual imagery to aid our grasp of space: We see Edwin Hubble's expanding universe by imagining the proverbial raisins in the bun moving apart from each other at a speed directly proportional to their distance apart when the bun began to cook. We can compare this image to the classic Ptolemaic universe of perfectly concentric circles, and our minds are set vibrating with intuitions of their vast difference. No one ever seems to have made a compelling image of the Bible's six-thousand-year-old universe measured in human lifetimes, so that there is no picture off of which we can bounce our new temporal vision. Nevertheless, there are verbal analogies aplenty for our new abyss of time. To take but one: If the initial condensation of the earth occurred on January 1, living things would have first appeared in the sea in May, and land plants and animals would have emerged in late November. The dinosaurs would have become extinct only on December 25, *Homo sapiens* would have first appeared at 11 P.M. on December 31, Rome would have ruled the Western world for the five seconds between 11:59:45 and 11:59:50, and three seconds before midnight Columbus would have discovered America.[1] Such temporal diagrams do not so much set our minds reverberating with a new awareness as send us reeling in defeat before something beyond our comprehension.

Another reason our grasp of the time revolution falls so far short of that of space is that, at least in its beginnings, it took so much longer to happen. It had no Copernicus and Galileo to abruptly set the new vision in motion. It was an evolution rather than a revolution, a slow, arduous struggle to make sense of discoveries about our planet and in the process to break with the temporal paradigm of the Bible and find a new one. It is a great, long, complicated tale which deserves to be told over and over again, but in the following pages we can only take the most cursory glimpse at a few of the key discoveries which began to bring the question of the age of the earth into focus. Ours will be an over-simplified and partial story—the barest bones of a convoluted saga of discovery. Yet our subject is *time*, and only relatively late in the game did time come into its own as a focus of interest. Until then, new knowledge about time was for the most part a disturbing and unwanted by-product of efforts to understand the processes underlying the visible structures of our planet.

To take the first step: we have René Descartes (1596–1650) to thank for dissolving the time barrier imposed by the belief in the world's impending End. He did it not by coming up with a new idea about time, but by introducing the idea that the physical world operates according to unchanging laws. In other words, if the time span of a pendulum's swing depended on the pendulum's length when Galileo discovered this law in the late sixteenth century, it depended on its length when the pyramids were built and will still depend on its length when we earthlings are living in biospheres on distant planets. Such an idea seems elementary and self-evident to us, yet one could not hold it and still believe that the world was decaying and would soon be put out of its misery by Divine fiat. If the same unchanging laws govern the physical world now and forever, then the world should go on indefinitely just as it is now. In fact, with the same laws in place, the earth *could not* be decaying.[2] God was not unseated by this new mechanistic vision; instead of ruling by intervention, God could now be seen as the prime mover who set in motion the universal laws of nature, which would then run on forever.

Remember that the first pendulum clock was built in 1657, just seven years after Descartes' death, and it was the *clock* which became the great metaphor for this new concept of the world. The words of the great physicist and chemist Robert Boyle bear repeating: The universe "is like a rare clock . . . where all things are so skillfully contrived, that the engine being once set a-moving, all things proceed according to the artificer's first design. . . ." We cannot be certain of Descartes' own views, but at one point he put it another way: "Give me matter and motion, and I will construct the universe." Whatever the belief about how the world began, as this mechanistic idea about what kept it going slowly diffused into society at large, it gradually put to rest fears of the world's imminent demise (and thus opened out future time). Nevertheless, it would be a long, long time before belief in the laws of nature would become the predominant way of thinking for the man in the street.

Once again, we are reminded of the many "times within Time": Descartes died in the same year Bishop Ussher announced the date and time of the world's advent. Despite its success in dispelling fears about the End, at least initially Descartes' concept of physical laws did little to discredit received opinion about the world's *Creation*. And there is something else to be aware of about this great idea. Although this one concept will have an over-arching impact on all the rest of our story, it says absolutely nothing about time per se. Descartes' vision of the world as a machine governed by physical laws was essentially a-historical, and we must look elsewhere for the first clues that the world might be something other than a static body created all-of-a-piece around 4,000 B.C.

Which brings us back to fossils. In the temporal realm, fossils are the closest thing we have to markers, the nearest analogy to the role played by the stars and planets in space. Although their presence before our eyes can't compete with the nightly sky show, people have recognized fossils as curious objects throughout history, and have had far more contact with them than city-bound people today may realize. In fact, we have been aware of fossils for at least as long as we have been studying the heavens. A 35,000-year-old Cro-Magnon fossil shell necklace has been found, and many neolithic sites contain fossils of all kinds, some of them pierced for wear as amulets. As we have said, a few Greeks—Anaximander, Pythagoras, Xenophanes, Herodotus—obtained the right answer when they realized the presence of marine fossils on inland mountains could only mean that the land had once been under water. This was a great clue both to the changes our planet has undergone and to the unfathomable amounts of time required for these changes to take place, but this clue was completely buried and forgotten during the following two thousand years.

Unfortunately, Aristotle was *not* among the clear-sighted Greeks when it came to fossils, and his great reputation was responsible for the long life of his notion that fossils were the result of "vaporous exhalations" from the earth. This theory of "spontaneous generation" continued throughout the Middle Ages, where it was joined by the idea that fossils were trial runs which the Creator decided not to bring to life. Another idea was that Satan had created fossils in a vain attempt to emulate God. And, of course, there was the theory that sea fossils had been carried to mountain tops by the biblical Deluge. Whatever the beliefs about their origins, mysterious powers were attributed to fossils—they were ground into powders to cure everything from nightmares to snakebites, and to neutralize poisons.

There were a few bright lights in all this fog. Leonardo da Vinci (1452–1519) refuted the theory that marine animals had been tossed onto mountain tops by the Deluge. One of his several closely reasoned arguments went like this:

> "If the deluge had had to carry shells three or four hundred miles from the sea, it would have carried the various kinds mixed and heaped up together; yet we see at such distances oysters all together, and conches and cuttle-fish, and all the other shells which live gregariously, all found together in death, while the solitary shells are found apart from one another, just as we may see them any day on the sea shore . . . "[3]

Without being able to explain precisely how, Leonardo nevertheless knew that "above the plains of Italy, where flocks of birds are flying today, fishes were once moving in large shoals." For Leonardo, the fish were not tossing about in the waters of a forty-day deluge, but were swimming in a normal, "permanent" ocean.

The person to take the first *two* giant steps toward finding an explanation for this extraordinary fact was a Dane named Niels Steensen (1638–1686), who changed his name to Steno when he went to Italy to serve as physician to the court of the Grand Duke of Tuscany. His first discovery involved the triangular-shaped stones known as *glossopetrae* (tongue stones), which were among the most common fossils. They were named by Pliny the Elder, who in the first century A.D. had suggested that they were petrified tongues that had fallen from the sky during eclipses of the moon. Until the eighteenth century, their supposed power to neutralize poison made tongue stones (in their own little tongue stone holders) part of the table-setting in proper European households.[4] When a large shark was caught near Livorno in 1666, Steno dissected it and discovered the perfect similarity between *glossopetrae* and the shark's teeth. He then wrote a treatise which showed, by word and picture, that *glossopetrae* were nothing other than fossilized shark's teeth. In essence, this simply brought the seventeenth century up to where the Greeks had been more than 2,000 years earlier.

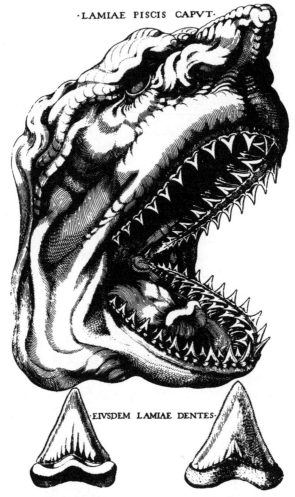

·LAMIAE PISCIS CAPVT·

·EIVSDEM LAMIAE DENTES·

27. Steno's drawing of the head of a shark, revealing that the teeth are identical to the mysterious *glossopetrae.*

Yet Steno went further than any Greek had gone when he suggested a mechanism by which the fossilized teeth "got into" the rocks in which they were embedded. (The biblical story of Creation caused people to believe that rocks had always been as hard as they are now, so the question of how fossils managed to get into rocks was a great puzzle. Indeed, many people didn't even get that far and refused to believe fossils were organic because, according to Genesis, the earth's solid rocks were created on the third day, and sea animals not until the fifth day.)[5] No one before Steno had ever suggested in print the revolutionary idea that rocks had perhaps not always been hard, but instead might have been soft sediments which later hardened around the marine animals and teeth.[6]

Steno also explained how the ancient animals might have been turned to stone (fossilized), but his contemporary Robert Hooke (1635–1703) actually studied fossils under microscopes, which he built according to van Leeuwenhoek's instructions. He also gave a good explanation of how the animal substance of fossils is slowly replaced by inorganic material. Thus, for all thinking people, fossils were no longer oddities that fell from the sky, but irrefutably long dead animals and plants. Although Steno seems to have accepted the Creation story unquestioningly ("And God created . . . every living creature that moveth"), Hooke saw extinction as the most probable explanation for why many fossils didn't look like any living animals. He suggested that the animals became extinct when countries sank below the sea or were raised up from the depths.[7] Thus, by the late seventeenth century the great issue of extinction of species had already been broached, and the question of *how long* the fossilized animals had been dead was there for the asking.

Steno's second great breakthrough was already augured in his brilliant insight that just because rocks are solid now does not mean that they have always been solid. He looked at the earth in a way that perhaps no human being had ever looked at it before, and he realized each layer of sedimentary rock had been laid down horizontally as silt, which eventually became compressed and hardened into solid rock. Each successive layer had to be younger than the one beneath it, and any breaks in a layer must have resulted from erosion after they had formed. If a sedimentary rock layer were tilted, it had somehow been uplifted after it originally solidified. "With a few simple principles, [Steno] transformed the earth's surface jumble into a legible archive."[8] As the word *archive* suggests,

Steno had discovered that the earth's history could be read from its strata, and so had found a "book" to set beside the Bible. This new "book" was the seed whose shoots would eventually force their way through the tough dogma of the Bible's static, six-thousand-year-old world to flower into a stupifying new temporal reality which we can know but can never truly comprehend.

But not yet. Steno's own life reveals the hold of the Bible on seventeenth century thought. He published his great treatise at the age of thirty-one, six years later he abandoned science, was ordained as a priest, and spent the remainder of his life as a zealous servant of the Catholic Church. Such an extreme reversal of direction may indicate that Steno couldn't handle the discontinuity with the biblical timeline already hinted at by his discoveries. Clearly, so many different layers of strata couldn't all have been laid down by a single Noachian Deluge, and thus the history of the earth appeared to be more complicated—and to have required more time—than described in Genesis. Once fossils of extinct species were clearly identified, God's creation of all species on the fifth and sixth days of Creation would also be called into question. This, too, would carry the implication of far more time than six thousand years.

It is impossible for us today to comprehend the situation of the early students of the earth. In the first place, they never set out to discover anything about time. They simply wanted to understand the physical processes of our planet. The discoveries about time dawned slowly and unbidden and only as a corollary of their other research. The idea that light could be shed on that intangible, abstract thing called *time* by looking into the depths of the solid earth is about as counter-intuitive as anything could be. Even today, if we stop to think about it, the connection seems truly weird. The great testament to the early students was that they saw it at all—they could never have been expected to grasp its awesome significance immediately. Although they couldn't ignore the disquieting implication that there was a lot more time involved than six thousand years, neither could they give up the only temporal paradigm they knew—and one which came with all the weight of scriptural authority. It was a profound dilemma, and it explains why geology moved forward so much more slowly than the other sciences which had their birth in the great seventeenth century.

For nearly a century the temporal dilemmas broached by Steno's discoveries were ignored. The earth was studied, fossils were collected, rock types were catalogued, but it wasn't until late in the eighteenth century that a theory attempted to explain how all this might have happened in something approximating the brief span allotted by the Scriptures. If you are short on time, the only way to make a lot happen is to make it happen quickly and violently. In order to explain the abrupt discontinuities between the different rock strata, the theory of *catastrophism* hypothesized a series of worldwide cataclysms in which floods, earthquakes, and all manner of violent displacement of land and sea destroyed the previous world and caused a new one to be created in its place. If catastrophism explained the appearance of the earth, it failed to find a natural cause for the catastrophes themselves. Most adherents of the theory had no problem with invoking divine intervention. In the nineteenth century, that intervention would again be invoked to explain the new fossils which appeared after each catastrophe: In each succeeding world, God was believed to have created new forms of life, so that it was these mini-Creations which explained the succession of different fossils in the strata. Clearly, by this time even those most devoted to the Bible's account realized that its one recorded deluge couldn't explain what the earth revealed.

The Bible's flood was thus seen as only the last in a long series of world-shattering catastrophes. And the Creation's "days" were now read metaphorically rather than literally, and thus as epochs encompassing indefinite periods of time. Although time had been stretched beyond six thousand years, the Bible's account was still involved in the explanation. By the turn of the nineteenth century, catastrophism had become the standard definition of the earth's history. Darwin himself may have even been behind the times when he set out on the

Beagle in 1831. By his own admission, at that time he believed in "the strict and literal truth of every word in the Bible."[9]

It is difficult to speak accurately about catastrophism because it has proven to be such a protean concept. Here we have spoken only of its early formulation. The catastrophists were in fact extremely empirical in their literal reading of the "book" of the strata. This book is a lurching, disjointed chronicle, with radically dissimilar rocks (and fossils) abutting each other. That catastrophes had occurred between these layers to cause their abrupt discontinuities was a most reasonable explanation. What the catastrophists did not realize was that at any particular outcrop huge stretches of time go unrecorded in the strata because their evidence at that location has been destroyed by episodes of heat and pressure due to mountain building and continental collisions. "A section of the crust at

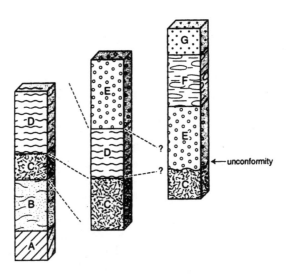

28. The relative time sequence of the stratigraphic record was pieced together from strata (and their distinctive fossils) found at different locations. In the rightmost column, note the unconformity between C and E, indicating a time gap in the record. All of D and part of C were eroded away in this location before E was deposited, but because D and the remainder of C were found in the place recorded by the middle column, they are known to have existed.

any one point corresponds to—say—an Egyptian chronicle containing only a few years at a time, taken at random from different dynasties."[10] It was only by painstakingly intercalating strata found in many different geographical regions that the complete stratigraphical record (known as the geological column) was finally built up.*

During the reign of catastrophism, John Harrison's watch found the longitude at sea, Captain Cook used it to map the islands and continents of the Pacific, and the timetables of England's mail coaches initiated the first efforts to coordinate time and distance on land. We were relating time to our turning planet as never before, but the vast majority of people still saw the earth's whole lifetime as within a few multiples of six thousand years. What often gets overlooked in this disparity is how *extremely* difficult it was—despite Steno's insight—to intuit the earth's past from the strata, and how that difficulty was hopelessly compounded when it came to trying to fathom the *time* that was involved.

* * *

Thanks to the Industrial Revolution, the "book" of the strata was becoming more and more open. In order to supply industry with the iron ore and coal it needed, mining operations escalated during the late eighteenth century. James Watts' improved steam engine helped to dredge water from the mines, and as more water was pumped out, the mine shafts could go deeper and deeper. These mines revealed strata of the earth never seen before. In turn, the iron ore and coal which they produced helped to build and fuel the steamboats—and later the railroads—which transported the materials of industry. Canals for steamboats required flat courses, as

---

*Nevertheless, today's scientists realize that natural catastrophes have cast a long shadow over the earth's history. Our planet's life has indeed been punctuated (and propelled) by catastrophes, the most significant of them arriving from outer space. To mention only a couple from a long list: the Mars-sized object that slammed into the earth soon after its formation and gouged out the material that became our moon, and the giant meteor that hit the gulf of Mexico sixty-five million years ago and whose atmospheric effects caused the extinction of the dinosaurs.

would the railroads a generation later, so hills were cut through to reveal yet more layers of the earth.[11] Never before had so many pages of the earth's "book" been laid wide open to view. (To this day, roadcuts made for superhighways remain a lodestone for geologists.)

It is interesting to contemplate the impact of the Industrial Revolution on both the time of day and on the slow, difficult process of trying to decipher the inscrutable language of the stones. When we think of the Industrial Revolution, the last thing we normally consider is its contribution to our ideas about time. And yet, as we saw earlier, the railroads had everything to do with quickening the pace of time in everyday life, and also with also by its coordination across space in the form of zone time. Would our awareness of the earth's eons have dawned when it did had not so many layers of our planet's crust been laid bare as a by-product of industry's forward thrust? Although the most exhaustive chapter in the earth's book, the Grand Canyon, had been discovered by Spanish explorers in 1540, they were no more aware of the possibility of reading the eons of time written in its strata than were the Indian tribes who had inhabited the Canyon for the previous four thousand years. At the turn of the nineteenth century, when minds in Europe and on the British Isles were churning with desire to make sense of it, the available pages were more often being hacked from the hills by man than being found ready-made by nature's processes.

* * *

Although the reader may not be aware of it, we are now standing at the edge of a precipice or, if you prefer to imagine it another way, on the farthest point of a continental shelf. One more step and we will plunge into an abyss so deep and chill that it will make the miles and miles of shelf seem like a cozy and familiar home. Once we fall off, we will never, ever be able to return.

Let's look for a moment at the person who will force us to take that one last terrifying step. There is a wonderful democracy about great ideas. They strike whom they please, rich or poor, healthy, or sickly. James Hutton's contemporary John Harrison was a self-taught clockmaker who scrounged for money all his life and still managed to build the clock that found the longitude. James Hutton (1726–1797) was an independently wealthy Scotsman who studied in Edinburgh, Paris, and Leyden, and knew both law and medicine. He ended up as a gentleman farmer, without a reason in the world to exert himself beyond the pleasant efforts that occupation required. And yet, at the age of forty-two, he gave up country life, moved to Edinburgh, joined the circle of James Watts, Adam Smith, and David Hume, and twenty years later gave the world its first revelation of "deep time,"[12] its first glimpse into our unimaginably long past.

It was a fact waiting to be discovered. It had waited since the dawn of human intelligence, apparently for a slender, aquiline-faced, eighteenth century Scotsman named James Hutton. It had strewn the earth with its clues—its places of exposed strata, its fossils, its processes of erosion and sedimentation, its volcanoes, earthquakes, and mountain chains—but no one until Hutton had followed the clues to their dark secret. (As we will see, there was one clue he died too soon to know. When it was found it would correct his one error, and we would finally have a sense of the time that we would one day be able to count.) Just as Copernicus had turned away from the "given" of an earth-centered cosmos and followed the clues which told him the sun was at the center, so Hutton simply gave up the catastrophists' struggle to hang onto the biblical timeframe by stretching it.

By jettisoning the need for a short timeframe, Hutton was able to explain the earth's changes in a manner that made sense in terms of the laws of nature. If the law governing the swing of Galileo's pendulum is the same now and forever, so, too, must be the laws which govern the earth's processes. *The processes we see occurring today must thus be the same processes that occurred in the past.* Given enough time, Hutton saw that the glacial-paced changes visible today could account for all of the transformations in rock strata and climate which have ever been or ever will be. If eons and eons of time were factored into the equation, the picture changed completely and there was no longer any need to hypothesize sudden, supernaturally-induced, world-destroying cataclysms. (It would not be until a generation later that William Whewell would call Hutton's theory *uniformitarianism*—an accurate if unwieldy term which has stuck to it ever since.)

By letting time expand exponentially, Hutton allowed the earth to be as obedient to the laws of nature as the planets circling overhead, but we will soon see that this did not yet mean that it could simply speak for itself. Does *anything* ever speak solely for itself, free of preconceptions on the part of the observer? Certainly nothing in this story has so far.

In 1788, Hutton published his theory in *The Transactions of the Royal Society of Edinburgh*, and in 1795 a vastly expanded version was published as *The Theory of the Earth*. In this work, Hutton also supplied a crucial ingredient missing from earlier explanations of geological change. These explanations had focused on the processes by which water wore down the earth by erosion, but they contained no concept of how the earth might be built back up again. This lack of any "concept of repair" buttressed the biblical chronology because it also indicated that the earth had to be relatively young. The mountains still towered above, and it wouldn't have taken so long for them to have eroded down to nothing.[13]

Until Hutton, most students of the earth believed that *all* rocks were formed by the wearing down, or decay, of the earth, primarily by the forces of water. Hutton realized that many crystalline rocks (like granite and basalt) did not crystallize out of water, but instead had an igneous origin. He correctly surmised that these "fiery" rocks were created when molten underground rock material, called magma, forced its way to the earth's surface when volcanoes erupted or through other vents for the forces of heat within the earth. With this insight, he had discovered how the earth built itself back up again.

When Hutton looked at the earth, he saw a perpetual motion machine forever cycling between erosion and uplift. Rivers and seas erode the land, and the loosened soil is ultimately carried to the ocean floors, where it is deposited and eventually consolidated into horizontal rock strata. The weight of ever more layers of strata generates heat and pressure, which causes the lower layers to melt. These melting sediments and the igneous magmas beneath them then expand and eventually burst forth to form new islands and continents, leaving the eroded old continents to become the beds of new oceans.[14] In their turn, these new continents begin to erode, and the whole cycle starts over again. Hutton had already shattered the biblical time barrier by making the past subservient to the same imperceptibly slow processes operating today. Now he had found the mechanism by which this glacial pace could go on forever and ever.

So, you may ask, are we now *finally* ready to move toward finding the true age of the earth? Unfortunately, we are not, and we are not because of a certain word that cropped up twice in the preceding paragraph: *cycle*. Hutton couched his great insight of time's vastness in terms of *cyclical time*. He believed that God began the world and might someday end it, but Hutton felt that such things were beyond our understanding. His focus was on what lay between those two poles, and there he saw an eternal cycle between decay and repair, with "no vestige of a beginning,—no prospect of an end." Hutton had given up one ancient paradigm only to embrace the other.

Furthermore, he seems to have embraced the cyclical paradigm with the same a priori faith that others had felt for biblical chronology. Stephen Jay Gould has effectively demolished the myth of Hutton the quintessential empirical observer by showing that his most important ideas materialized in his armchair, and that he only went into the field to corroborate them. The stones still have a way to go before they can speak for themselves, and what we said about the aid given by the Industrial Revolution to geological knowledge does not apply to its most prescient student. Although Hutton compared the cycles of the earth to the revolving planets of Newton's cosmos, they were of course the same planets which were the impetus for the ancient world's conception of time as cyclical. Hutton hoped that "the apparent messiness of [the earth's] complex history could be ordered as a stately

cycle of strictly repeating events, [so that] the making and un-
making of continents might become as lawlike as the revolution of planets."[15]

We said at the outset that the journey toward the marriage of the ancient world's awareness of "the inhuman workings of water on stone" with the Judeo–Christian intuition of
the linear, directional nature of time would be long and tortu-
ous, and we have just proved the point. The plan had been to
end this chapter with the one clue that Hutton missed, and
thus to consecrate that marriage, but we'll save it for the next
for the sake of another form of unity. By omitting it here, we
have lent to this chapter and the preceding one a certain
palindromic symmetry. In the previous chapter, we began
with cyclical time and ended with linear, and here we have
done the reverse. We seem to have gotten not one whit closer
to our goal of discovering the age of the earth, which only
proves how *extremely* difficult it is to fathom time. If the
earth's clues are lying there for the taking, time itself is
sphinx-like. It doesn't send out clues—its clues can only be
found in things not itself. But it is powerfully *real*, and in the
end, by the slow increments gained on this zigzag itinerary, it
will ultimately give up its secret.

Some progress has already been made, because the symmetry between the two chapters is not perfect. Hutton's geological cycles, imprinted on strata, mountain chain, and sea,
bear no more relation to the mentally-elaborated cycles of the
ancient world than would Rutherford's atom to Democritus'.
Because they are abstract, the ancient time cycles don't send
a chill down the spine as Hutton's do—they lack the almost
tangible, physical reality that Hutton gave to time.

The second reason for holding off the clue is out of defer-
ence to the man himself. He should not be squeezed to make
room for anything. It is a travesty that Hutton's name is virtually unknown to the general public, while the stars of the
spatial revolution, like Copernicus, shine brightly in the
minds of children. Finally, two and a half centuries after
Copernicus had displaced humankind from the center of
space, our world of time was no longer something that had

been created just for us. By puncturing the boundaries of time, James Hutton gave us "geology's most distinctive and transforming contribution to human thought."[16] It does not matter that he could not *count* the eons of time he had discovered. It does not even matter that he saw them as cyclical and thus did not care to count them. After all, biblical scholars could use human lifetimes to count their six thousand years very well, but their basic assumption turned out to be fiction. It is enough, more than enough, that Hutton had discovered endless, endless time, time past all conceiving. Hutton had found the real thing—the matrix that eventually would be counted. To use an analogy from his own insight, *given enough time* it could be accomplished.

Chapter XIX

The Battle of the Geologists and the Physicists

In civil history, we consult titles, we research medals, we decipher ancient inscriptions. . . . Similarly, in natural history, it is necessary to excavate the archives of the world, to draw old monuments from the entrails of the earth. . . . This is the only way . . . to place a certain number of milestones on the eternal route of time.
Georges Leclerc, Comte de Buffon (1707–1778)

Whenever you can, count.
Sir Francis Galton (1822–1911)

In 1769, in a little town in Oxfordshire, England, a child with the very ordinary name of William Smith was born into the poor family of a village blacksmith. This unpropitious beginning was compounded when, at the age of seven, William lost his father and was sent to live with an uncle. He received rudimentary village schooling, but mostly he roamed his uncle's farm collecting the fossils that were so abundant in the rocks of the Cotswold hills. His uncle didn't approve of William's intense interest in these odd objects, but then he had no way of knowing that a great gift to knowledge lay like a germinating seed deep in the mind of his strange nephew.

When he grew older, William Smith taught himself surveying from books he bought with his small savings, and at the age of eighteen he was apprenticed to a surveyor of the local parish. He then proceeded to teach himself geology, and when he was twenty-four he went to work for the company that was excavating the Somerset Coal Canal in the south of England. This was before the steam locomotive, and canal building was at its height. The companies building the canals to transport coal needed surveyors to help them find the coal deposits worth mining, as well as to determine the best courses for the canals. This job gave Smith an opportunity to study the fresh rock outcrops created by the newly-dug canal. He later worked on similar jobs across the length and breadth of England, all the while studying the newly revealed strata and collecting all the fossils he could find. Recall that fast mail coaches had been the first instigators of the time-telling reforms which would eventually lead to zone time. They played a role in this story, too, because Smith used them to travel as much as ten thousand miles per year. In 1815, he published the first modern geological map, "A Map of the Strata of England and Wales with a Part of Scotland," a map so meticulously researched that it can still be used today.

In 1831, when the lowly-born Smith was finally recognized by the wealthy, class-conscious members of the Geological Society of London as the "father of English geology," it was not only for his maps, but also for something even more important. Ever since people had begun to catalogue the layers of strata in particular outcrops, there had been the hope that these could somehow be used to calculate geological time. But as more and more accumulations of strata were catalogued in more and more places, it became clear that the sequences of rocks sometimes differed from region to region, and that no rock type was ever going to become a reliable time marker throughout the world. Even without the problem of regional differences, rocks present a difficulty as unique time markers. Quartz is quartz—a silicon ion surrounded by four oxygen ions—there's no difference at all between two-million-year-old Pleistocene quartz and Cambrian quartz created over 500 million years ago.[1]

As he collected fossils from strata throughout England, Smith began to see that the fossils told a different story from the rocks. Particularly in the younger strata, the rocks were often so similar that he had trouble distinguishing the strata, but he never had trouble telling the fossils apart. While rock between two consistent strata might in one place be shale and in another sandstone, the fossils in that shale or sandstone were always the same. Some fossils endured through so many millions of years that they appear in many strata, but others occur only in a few strata, and a few species had their births and extinctions within one particular stratum. Fossils are thus identifying markers for particular periods in the earth's history.

Not only could Smith identify rock strata by the fossils they contained, but he could also see a pattern emerging: certain fossils always appear in more ancient sediments, while others begin to be seen as the strata become more recent. By following the fossils, Smith was able to put all the strata of England's earth into *relative* temporal sequence. About the same time, Georges Cuvier made the same discovery while studying the rocks around Paris. Soon it was realized that this *principle of faunal succession* was valid not only in England or France, but virtually *everywhere*. It was actually a principle of floral succession as well, because plants showed the same transformation through time as did animals (fauna). Limestone may be found in the Cambrian or—300 million years later—in the Jurassic strata, but a trilobite—the ubiquitous marine arthropod which had its birth in the Cambrian—will never be found in Jurassic strata, nor a dinosaur in the Cambrian. By themselves, the rocks could never give an unambiguous story of the earth's history—only the ever-onward march of the fossils was unerring. By the mid nineteenth century, the fossil sequence had helped to put into place most of the periods of our familiar geological column.

THE GEOLOGICAL COLUMN

Era	Period	Epoch	Age (millions of years)	First Life Forms
		Holocene	0.01	
	Quaternary			
		Pleistocene	3	Man
Cenozoic		Pliocene	11	Mastodons
		Miocene	26	Saber-toothed tigers
	Tertiary	Oligocene	37	
		Eocene	54	Whales
		Paleocene	65	Horses Alligators
	Cretaceous		135	
				Birds
Mesozoic	Jurassic		210	Mammals Dinosaurs
	Triassic		250	
	Permian		280	Reptiles
	Pennsylvanian		310	
	Carboniferous			
Paleozoic	Mississippian		345	Amphibians Insects
				Sharks
	Devonian		400	
	Silurian		435	Land plants
	Ordovician		500	Fish
	Cambrian		570	Sea plants Shelled animals

29. By following the fossil sequence and the changes in the rock strata, the eras and periods of the column were slowly pieced together during the late eighteenth century and first half of the nineteenth, but it was not until the discovery of the radioactive clocks in the twentieth century that their ages could begin to be determined.

This was the clue that Hutton had not lived to see (although his fierce commitment to a cyclical earth might well have prevented him from accepting its implications even if he had been aware of it). The fossils, of course, revealed nothing whatsoever about *absolute* time; there was no way to attach a date to any of them. They only marked what came before or after. When they were all lined up, however, they showed something else: as the strata ascended and thus became more recent in time, the fossils within them tended to become more complex, and once a fossil disappeared from the strata, it never reappeared. This cannot happen if time is cyclical. It only happens if time is moving in a direction.

At the time of its publication, Hutton's theory of an endlessly repeating cyclical earth had not gained much of a following. Catastrophism was too entrenched, Hutton's view was too radically removed from the biblical account, and his writing was in any event simply too opaque to gain a wide readership. It was Charles Lyell's (1797–1875) three volume *Principles of Geology*, first published in the 1830s, which finally presented Hutton's cyclical, uniformitarian vision in a way that gained it considerable acceptance. But at some point Lyell and his followers were going to have to come to terms with a fossil record that changed irreversibly as it moved ever onward through time, and from which species periodically disappeared forever.

As a Creationist in the late eighteenth century, Thomas Jefferson had loathed the idea of extinction and had instructed the explorers of the American West to keep their eyes open for woolly mammoths. Although their fossilized bones had already been discovered, Jefferson reasoned that this did not prove their extinction, but merely indicated that they weren't now inhabiting areas where people would come across them. Analogously, the uniformitarian Lyell now suggested that the apparent irreversible changes in species through time were just that. When milder climates cycled back in future eons, "the huge iguanodon might reappear in the woods, and the ichthyosaur in the sea, while the pterodactyle might flit again through the umbrageous groves of tree-ferns."[2] And once the strata of the Paleozoic era could be more deeply studied, Lyell believed that later animals such as mammals would surely be found there. But as the scrutiny of Paleozoic strata extended into Eastern Europe and North America, not a single Paleozoic mammal was ever found. Finally, in the tenth edition of his *Principles of Geology*, published in 1866, Lyell capitulated, and admitted that the changes in organic life revealed by the fossils indicated their linear progression through time.[3]

Others had believed the significance of the fossils several decades before Lyell, so that during the first half of the nineteenth century something of the vision of earth's history which we still hold today had already emerged in embryo. It required then, as it does now, the marriage of Hutton's vast eons to a linear concept of time (now revealed by the fossil record instead of by the Bible's sequence of human lives). The eternal cycle of Hutton's eons had been broken, and "deep time" had been stretched out into a line, the end of which was lost in an unfathomable abyss. The great difference between this vision and our own was that it had no starting point for its linear time. The possibility that there was no starting point, that the earth's time was infinite, could not be ruled out. The first efforts to find a starting point would come from people who looked at the earth from a very different vantage point than the geologists.

So far we have only looked at the earth beneath our feet. By studying the actual, physical crust of the earth, the geologists had come to realize that the earth had a history just as vital and ultimately decipherable as the history of the people who walked upon it. The Bible's linear time referred only to people; the earth had merely been the static backdrop against which their pageant was played out. During the eighteenth century the idea had gradually dawned that time and the changes it wrought were just as integral to understanding the earth, and nature, as they were to understanding human history. With the publications of Charles Darwin's (1809–1882) *Origin of Species* (1859) and *Descent of Man* (1871), and their elucidation of the mechanisms driving the changes in all living things, the ancient separation of natural history from human history would come to an end, the tables would be turned, and humankind would at last become part of nature.

By looking only at the ongoing developmental processes of the earth's crust, the geologists had stepped in midstream— none of them was concerned with going back to the beginning and questioning the Genesis account of our planet's origins. Yet the idea of our planet's birth is as much a part of our own notion of the earth's history as is the image of a fossil record extending into unimaginable depths of time. We must now backtrack a bit to take a look at the physicists, who followed an entirely different route to arrive at the same conclusion as the geologists that the earth was far older than six thousand years.

What if, instead of looking down at the earth beneath your feet, you instead pulled back and ask yourself how this whole giant globe of the earth could originally have formed? Clearly, this line of questioning completely disregards the Genesis account, and yet the conflict seemed to present less of a problem to the physicists than it had to the geologists. Already in the seventeenth century, Descartes sketched a theory of how the universe could have evolved by natural processes. Although his theory turned out to be wrong (it involved the idea that the earth was a burnt-out sun) it was the source of a whole succession of ideas about how the cosmos could have evolved.[4] In the year 1796 both Pierre-Simon Laplace (1749–1827) and Immanuel Kant (1724–1804), quite independently, arrived at the nebular hypothesis for the origin of our solar system: early in the sun's history, it was surrounded by a gaseous atmosphere which rotated around it. As time went by, large areas of this atmosphere coalesced, and eventually the force of gravity caused planets to form at the centers of these coalesced masses. The moons of the planets similarly formed from the remaining material around the planets. This hypothesis explains why all the planets orbit in nearly the same plane and move in the same direction. An updated version of the nebular hypothesis is the reigning theory of the solar system's origin today.

For Laplace, this theory referred only to the birth of the solar system, and one of its virtues for him was that it insured maximum stability to the system thereafter, so that it could then cycle on forever in accordance with the laws of nature.[5] Like Hutton, Laplace seemed to believe that any change beyond that of dynamic cyclicity was precluded by the world's obedience to the laws of nature. Kant knew otherwise—constancy of laws does not preclude directional change—and he thus found a very different meaning in the same theory.

If the earth had traditionally been seen as exempt from the changes which time brings to humankind, how much more so had the heavens! The seemingly eternal cycling of the heavens had been the basis of our first theory of time, and paradoxically, our first awareness of the world's endless eons came to us in part because Hutton so desired to give to the earth's changes the sanction of eternal order which Newton's laws had given to the sky above. And yet half a century before Hutton, Kant had already displaced the age-old concept of a static, cycling heaven with that of one which changes through time, which *evolves*. In his *General History of Nature and Theory of the Heavens* (1755), Kant not only explained the Milky Way galaxy as a kind of vastly enlarged solar system, but he realized that the "luminous patches" caught by telescopes in the night sky were in fact other galaxies like the Milky Way. Kant believed that the universe had a beginning, but that it would continue to evolve forever: "Creation is not the work of a moment . . . Millions and whole myriads of millions of centuries will flow on, during which always new Worlds and systems of Worlds will be formed, one after another. . . ."[6]

The person who brought this new linear cosmic time down to earth, so to speak, was the great Georges Leclerc, Comte de Buffon (1707–1778). Someone once described Buffon as being more like a committee than an individual person. When he wasn't in Paris participating in the salons of the Enlightenment, he was successfully running his large estate in Burgundy. And these were the least of his activities. During the thirty-six years from 1749–1785, Buffon published a volume per year of his best selling *Histoire Naturelle*, in which he discussed in loving detail virtually every known plant, animal, mineral, and climate of the natural world. His *Epochs of Nature* (1779) was an early, brilliant effort to explain Creation in naturalistic rather than biblical terms. Buffon was as many-sided in his scientific work as he was in his personal life, and he stood almost alone in his ability to look at the earth both as a naturalist and a physicist.

Most naturalists (and geologists) stopped their explanations at the point of Creation. "In the beginning God created the heaven and the earth" was a terra incognita which was not subjected to the metaphorical interpretations given to God's other creative acts. Not only did Buffon not stop where others had stopped, but he didn't bother with metaphors either. He simply put on the hat of physicist and speculated that the earth was formed when a comet collided with the sun and caused hot gases and liquid to spew out and eventually form the planets and their moons. (This catastrophist theory has been the only serious rival of the nebular hypothesis of the solar system's origin, but the latter now appears to be correct.) Buffon thus believed that the earth began white-hot and molten. It clearly wasn't in that state now, so if the rate of its cooling could be determined, the age of the earth could be known.

Buffon set up a laboratory in his cellar in order to minimize temperature fluctuations caused by the sun. He persuaded the foundry workers on his estate to make ten iron balls from an inch in diameter to five inches, increasing by half-inch increments. When he heated them all to white heat and then measured the time it took for each of them to return to room temperature, he saw that the relationship between diameter and cooling time was approximately linear. He then had more balls constructed which contained materials closer to what he believed was the actual composition of the earth. He made corrections for the delaying effects of the sun's heat, as well as for major events he believed had occurred in the earth's history. He extrapolated the rates of cooling he found in his little balls to the rate for a ball the size of the earth, and concluded that the earth was now 74,832 years old!

We may smile at a precision in the result reminiscent of Bishop Ussher a hundred years earlier, but there was nothing else about this experiment reminiscent of Bishop Ussher. Although Buffon's figure for the earth's age was only about thirteen times greater than Ussher's, it was in every other respect light-years beyond it. Its quantitation was based on physical laws rather than on human lifetimes, and what it quantitated was not the age of *everything*, but rather the age of that single, unique third planet out from the sun.

The geologists had found no way to quantitate the eons of time they had discovered. Buffon's experiment offered nothing like Hutton's awesome new vision of time, and yet it was the first step on the other road which would eventually dovetail with that taken by the geologists to bring us to the knowledge of the earth's history we have today. Today we have Hutton's eons, but they move in a direction and they are *quantitated*. Without that quantitation, our awareness of them would be nearly as vague and abstract as were the cycling eons to the ancient pagan cultures. It is truly half of our own vision.

The 4.5 billion years we know today is six thousand times longer than Buffon's 74,832 years. The story of how we got from there to here in two hundred years belongs primarily to the physicists and the other time counters rather than to the geologists and naturalists. And yet, sitting in the middle of the story throughout the nineteenth century would be the colossal mountain of time discovered by Hutton and absolutely *required* for Darwin's theory of natural selection to be valid. As the physicists went on with their far-too-small calculations, Hutton's and Darwin's intuited but unquantitated (and thus unproven) mountain of time would not go away. Finally, at the turn of the twentieth century, an as yet unimagined branch of physics—*nuclear* physics—would begin to prove all the earlier time counters wrong. The mountain of time would explode into the sky beyond the farthest Everest conceived of in the nineteenth century, and man-years of toil and thought by brilliant people would go down the tubes.

Upwards of two hundred different calculations were made for the age of *everything* using human lifetimes as the unit of measurement. Buffon's figure based on the rate of cooling of the earth stands virtually alone in the eighteenth century, and it would not be until the second half of the nineteenth century that another great spate of calculations about the earth's age would appear. In essence, these would result from the first crude clocks for determining the age of the earth. Their inventors picked one aspect of the physical environment and tried to measure how much it had changed since the earth began, much as the Egyptians picked out the sun and stuck a stick in the ground to measure its change throughout the day. As we will see when we take a cursory look at a few of these new clocks, measuring out the pieces of an already-known pie is a far easier task than guessing at its ingredients at the same time you are trying to cut it, and the Egyptian gnomon was a veritable atomic clock in its accuracy by comparison to these first "machines" for measuring the age of the earth.

Recall that in our first chapter we said that the earth's day is constantly lengthening because the moon's gravitational pull on the earth—which causes the tides—creates friction between the oceans and the sea floor, thus causing the earth's rate of rotation to slow down. Assuming an original three to five hour period for the earth's rotation, George Darwin, Charles' second son, calculated that it should take fifty-six million years to reach our present twenty-four-hour day. Among several problems with Darwin's calculations was his erroneous assumption that the rate at which the earth's energy is dissipating is constant. We now know that the shapes of the oceans (and the amount of shallow seas) have changed enormously during the earth's history. Friction is much greater between the oceans and shallow bottoms than it is in areas of deep sea, so that there have been great fluctuations in the rate of the earth's energy loss. Darwin admitted that there was a high degree of speculation involved in his clock, and he would surely not have been surprised at today's conclusion that it will never be viable as a means of determining the age of the earth.[7]

The "salt clock" was based on chemistry rather than physics, and it was first suggested by Edmund Halley—better known for the comet which bears his name—in the early eighteenth century. It was resurrected in the nineteenth century, and was based on the idea that the oceans had existed almost as long as the earth, and that they began as essentially fresh water and slowly accumulated their salt content as rivers carried salt eroded from crystalline rocks into the sea. Unlike rivers, oceans have no outlets, so that their salt content was believed to slowly increase year by year. By calculating the salt content of the present-day ocean and the rate at which it was increasing, the age of the earth could be learned. The most thorough-going effort to use this clock was that of John Joly (1857–1933), who in 1899 announced that the age of the earth was approximately eighty-nine million years.

Aside from the wrong assumptions of a constant rate of sodium influx into the ocean and of there being only a tiny amount of sodium in the original ocean, Joly's great error was in assuming that sodium remained and accumulated in the ocean once it arrived. Today we know that sodium leaves the ocean at about the same rate that it enters it, both by evaporation and by incorporation into sediments on the ocean floor. Clearly, all the tinkering in the world could never make the salt clock tell the age of the earth.

Although initially quite hostile to the incursion of the physicists onto what they considered to be their turf, the geologists themselves came up with a clock based on the rate of sediment accumulation. In fact, so enamored of it did many of them become that it was second only to the biblical "lifetime" method in the total number of estimates for the age of the earth which it produced. Again, a complicated, much-corrected procedure can be capsuled in a deceptively simple formula: calculate the total estimated thickness of all layers of sedimentary rock and divide the result by the estimated rate of sedimentary deposition per year, and you will obtain an estimate of the time since sediments began to accumulate (which was roughly equated with the age of the earth). Despite laborious efforts to distinguish between varying deposition rates of different rock types, the word "estimate" tells the tale. As with the salt clock, uniform rates were incorrectly assumed. We know now that there is simply no such thing as an average rate of deposition for any geological time period. But the truly impassable barrier to this method was that, at best, it could only pretend to find an accurate measurement for the fossil-bearing sedimentary rocks comprising the 550 million years of the geological column. The remaining eighty-seven percent of the earth's lifetime is completely intractable to dating by sediment accumulation because most of its rocks formed by the cooling of molten magma, and nearly all of its sedimentary rocks were long since transformed from their sedimentary state by the forces of heat and pressure. The jumbled forms of these ancient rocks bear no resemblance to the layer cake appearance of the geological column's sedimentary rocks, and we can never ever attempt to date them by this method.

We don't really need to look at another of these early clocks to prove the point of their inaccuracy, and yet we must take a brief look at the one which led the "clockmakers" into a pitched battle with the geologists and naturalists of the Hutton, Lyell, and Darwin school right at the dawn of the twentieth century.

William Thomson (1824–1907) entered the University of Glasgow at age ten, and at the age of twenty-two was appointed Professor of Natural Philosophy there. During his lifetime he published over six hundred scientific papers and obtained dozens of patents. He was elected president of the Royal Society for five terms and was raised to the peerage (i.e., he became Lord Kelvin) in 1882. He was probably the most honored British scientist in all history. Thus, when Lord Kelvin spoke, people listened. One of the subjects he spoke about was the determination of the age of the earth based on the "cooling earth clock," the same clock that Buffon had invented nearly a hundred years earlier.

Lord Kelvin's clock was far more sophisticated than Buffon's ball experiment: Kelvin looked at the sun as well as the earth through the lens of thermodynamics, and concluded that both were irrevocably cooling. The second law of thermodynamics requires that whenever energy is changed from one form to another, a small amount of the energy is lost because it is dissipated as heat. The second law thus *requires* change to occur. In fact, it requires that the world *run down*.

So far from being a-historical (as most laws of nature seem to be), the second law of thermodynamics actually makes Hutton's idea of the earth as a forever-cycling perpetual motion machine impossible. Just as the ever more complex record of the fossils says that time has a direction, so too does the second law of thermodynamics. Kelvin's sun and earth were natural machines that were slowly running down (cooling) entirely in obedience to a law of nature.

Like the inventors of the other early clocks, Kelvin used unreliable estimates and then applied sound mathematics to those estimates. He believed that the sun's heat came from gravitational contraction, a process which clearly could not go on forever. Using estimates of the sun's current temperature and of the rate at which it was decreasing, Kelvin estimated that the sun had been shining for perhaps 100 million years and certainly not for more than 500 million years. Similarly, using figures for the initial temperature of the earth, the thermal conductivity of its rocks, and the current thermal gradient of the rocks in the earth's crust, in 1862 he calculated that the earth was about ninety-eight million years old. In 1897, he revised the figure downward to between twenty and forty million years. Because it was Kelvin, people were persuaded, and did not heed the warning of one critic, T.C. Chamberlain, who pointed out that "there is perhaps no beguilement more insidious and dangerous than an elaborate and elegant mathematical process built upon unfortified premises."[8]

Not only were Kelvin's premises estimates, but they were based solely on the primordial heat left over from the earth's formation. Kelvin knew nothing of the heat generated by radioactivity, which is ongoing and probably the single most important source of the earth's heat. Today we know that the earth's thermal history is so complicated and our knowledge of it doomed to such inadequacy, that it is highly unlikely that any form of cooling earth clock will ever be able to tell us the age of the earth.

Although it failed as a clock and is far from the idea of the earth we hold today, the cooling earth envisioned by Kelvin and Buffon nevertheless gave us our first awareness of the most sorrowful aspect of our own view of our planet: just as it once had a birth, so it will also one day have a death. And long before our planet's death, life on earth will have become impossible. Buffon calculated that in another 93,291 years the earth would have become too cold for life to exist upon it. Kelvin also knew that "within a finite period of time the earth must have been, and within a finite period of time to come the earth must again be, unfit for the habitation of man as at present constituted."[9] Today we know that instead of 93,291 more years, life probably has another 1.5 billion years to go on our planet, and we also know that it will eventually become impossible not because the earth has cooled too much but because the sun has heated up and expanded, evaporating the earth's oceans and transforming its atmosphere.[10] But the first glimmer of this desolation nevertheless came to us because Buffon and Kelvin looked at the earth as physicists and in the process tried to determine its age.

Even if we forget about the factors that were as yet unknown, none of these early clocks would have been able to measure *anything* accurately because in every case the *rate* of change they were measuring turned out to be variable. Their inventors clearly knew how crucial a constant rate was because they made a point of *assuming* it. They knew that this is the heart of any form of clock—the story of the traditional clock in the previous section is essentially a chronicle of the ever-improving uniformity of the period to be counted—from the pendulum's swing to the quartz crystal's oscillation to the cesium atom's frequency. By comparison, the clocks so far produced by the earth are so woebegone as to make the earliest mechanical clocks, with their losses of up to an hour a day, look like paragons of accuracy. How could this huge, turning planet with its rocky crust, molten interior, and complex thermodynamics ever be expected to produce a clock that could determine its age? It couldn't be—the miracle was that it did.

But not soon enough for Charles Darwin. Darwin died at the age of 73 in 1882, more than two decades too soon to learn about a new clock which would ultimately stand alone in its accuracy and finally corroborate what Darwin knew but could not prove: the earth is ancient past all conceiving. When he developed his theory of evolution by natural selection, Darwin had at the forefront of his mind the eons discovered by Hutton and substantiated by Lyell, whose book he took with him on the *Beagle*. He could not have arrived at his great theory had the idea of near infinite geological time not already been in his mind, and, without it, his theory could not possibly be correct.

By this time, God's Word was not enough for people anymore. They demanded *proof* of things before they would proffer their belief. Remember that it had taken the hard evidence of the linear fossil record to finally convince Lyell of the direction of time, and the hard evidence of the many layers of sedimentary strata to convince the catastrophists that one biblical deluge could not have caused them all. Although Kelvin's short time frame had many critics, they could not prove that 100 million years was not *nearly* long enough, and in the absence of hard evidence, many people's idea of proof was skewed toward mathematics. In reality, Darwin's theory with its awesome time requirements was as near to proof as the mathematics based on guesswork of the physicists. Yet people didn't see it that way; the physicists' *numbers* had the ring of truth that Darwin's theory lacked.

It was the problem of "fit" all over again. Just as the early geological findings wouldn't fit the biblical time scheme, so now geological/evolutionary time would not fit the calculations of the physicists. It was a case of two different views and two different methods at loggerheads.

In his letters, Darwin showed how troubled he was about Kelvin's short timescale. He wrote to Wallace: "I have not as yet been able to digest the fundamental notion of the shortened age of the sun and earth."[11] Darwin knew that both he and Kelvin could not be right. It is heartbreaking to think that Darwin died with this worry on his mind, but time, in its indifference, would not stop even for him.

Chapter XX

The Clock That Worked

*It is not in the premise that reality
Is a solid. It may be a shade that traverses
A dust, a force that traverses a shade.*

Wallace Stevens
An Ordinary Evening in New Haven

Although nobody realized it, a year before Lord Kelvin gave his famous speech diminishing the earth's age to 20–40 million years, the clue which would disprove him had already been discovered. It would dispell forever the disquieting aura of uncertainty which hovered over the thinkers of the late nineteenth century as they stared at the disparity between the eons of time required by Darwin and Lyell, and the calculations of Lord Kelvin. The year was 1896 and the place was a shed at the far end of the Jardin des Plantes in Paris. No one had any idea that this would be the birthplace of the answer to the great question of the age of the earth, least of all the man puttering quietly in the shed.

Like his father and grandfather before him, Henri Becquerel (1852–1908) studied luminescent materials in order to learn what made them glow after they had been exposed to sunlight. Excited by Wilhelm Roentgen's (1845–1923) discovery of x-rays in 1895 and intrigued by the fact that they caused luminescence in materials they came in contact with, Becquerel wondered whether the luminescent materials might somehow be the *source* of the x-rays. He set out to test this rather odd hypothesis using his own luminescent chemicals. First he wrapped photographic film in black paper (which sunlight can't penetrate), and then he put a crystal of a luminescent chemical on top of the paper and let them both sit in sunlight. After several chemicals gave no response, he tried one which happened to be a uranium salt. This time the film fogged, which meant that some kind of penetrating rays had to be coming from the uranium.

Although it was odd that the other chemicals hadn't caused the film to fog, Becquerel still assumed a causal relationship leading from sunlight to luminescence to the giving off of x-ray-like rays. He tried to do a second experiment to corroborate his results, but the sun didn't shine and thus the uranium didn't luminesce. He put the photographic film, with the uranium salts on top of it, in a drawer while he waited for the sun to shine. It was February in Paris, so he waited for days and days. We will never know what went through Becquerel's mind when, after days of waiting and still no sun, he decided to go ahead and develop the film anyway. On the face of it, it was an absurd thing to do—there should be nothing on the photographic plates. Nevertheless, he did it, and what he found was that the film was as strongly fogged as before. Whatever was coming out of the uranium crystals didn't require sunlight to make it happen! He soon verified that it didn't have anything to do with luminescence either. In some states uranium will luminesce and in others it will not, and Becquerel ran his experiment again using both. It didn't make the slightest bit of difference—the film was strongly fogged in both cases. There could be only one explanation: neither sunlight nor luminescence had anything to do with the rays. The uranium *itself* was the source of the mysterious penetrating rays. And it soon became clear that these rays were something quite different from x-rays because they were far too weak to make pictures of the insides of bodies.

Although it wouldn't be given that name until Marie and Pierre Curie christened it in a 1898 paper, what Becquerel had discovered was *radioactivity*, the spontaneous emission of subatomic particles by the atoms of unstable elements. The world did *not* stop in its tracks. Becquerel himself seemed to lose interest, and Marie Curie (1867–1934) chose to study the phenomenon for her PhD thesis in part because no one else seemed to be taking much interest. All the attention was focused on x-rays, which offered the high drama of pictures of things turned inside out, and which almost immediately began to be used for medical purposes. They were also easily produced with a vacuum tube and a high-voltage coil. In contrast, uranium was enormously difficult to obtain, and its only previous claim to fame had been its use as a coloring agent in ceramic glazes. In addition, because Becquerel's rays had some resemblance to x-rays, they were seen as somehow related. "It would take fresh eyes, and more careful quantitative methods, to establish that uranium rays were a part of another phenomenon altogether. X-rays might reveal hidden bones. Uranium rays were pointing the way to an understanding of the building blocks of all matter."[1]

Those fresh eyes belonged above all to Marie and Pierre Curie (1859–1906) and to Ernest Rutherford (1871–1937). Within less than a decade, their discoveries would knock out Kelvin's short chronology for the earth with a one–two punch, and Darwin's intuition of the earth's eons would at last be vindicated.

The first punch came with the discovery that radioactivity is an exothermic process involving the release of enormous amounts of heat. Marie Curie's heroic four-year-long struggle to isolate two other radioactive elements, polonium and radium, revealed that radioactivity was not just an oddity of uranium, but was a more general phenomenon. And in the process of isolating these new elements, it also became clear that the kinetic energy of the radioactive rays was being transformed into heat energy. When the radioactive rays collide with molecules in the air (or whatever medium surrounds them), they are absorbed by those molecules and their kinetic energy is transformed into heat. Just how much heat was revealed in a paper published by Pierre Curie and Albert Laborde in 1903: one gram of radium could heat a gram of water from freezing to boiling in an hour's time!

Well before this quantification of the heat was made, the mere fact of so much energy was a source of bewilderment. Where could it be coming from? Marie Curie expressed the problem clearly in 1900:

"When we observe the production of . . . Roentgen rays, we ourselves are furnishing . . . the electric energy . . . But at the time of the uranic [radioactive] emission . . . no change occurs in this material which radiates the energy . . . in a continuous fashion. The uranium shows no appreciable change of state . . . it remains, in appearance at least, the same as ever, the source of the energy it discharges remains undetectable."[2]

The words "in appearance at least" are crucial, because of course the uranium atoms *were* changing, but it would take just a little while longer to figure out what was going on. At the time, two incorrect hypotheses were advanced by the Curies and others: the first conjectured that radioactivity represented an exception to the first law of thermodynamics, which dictates that energy can be changed from one form to another, but can't be created or destroyed. Because no change could be detected, it was not unreasonable to ask whether radioactivity might actually involve the *creation* of energy. Alternatively, it was suggested that the radioactive materials might somehow be absorbing energy from the atmosphere.[3]

The latter hypothesis was tested by two German school teachers and amateur scientists named Julius Elster and Hans Geitel. They buried a radioactive substance deep in the Harz mountains and another 850 meters down in a mine shaft, assuming that without exposure to the atmosphere they would lose radioactivity. It didn't make the slightest bit of difference, and Elster and Geitel correctly concluded that radioactive substances had to be producing their heat energy themselves and not picking it up from the surrounding air.

Many other amateur scientists joined Elster and Geitel in outdoor experiments, and soon it was discovered that traces of radioactivity are all around us in soil, rain, ground water, snow, and rocks. Radioactive mist was even found at the base of Niagara Falls.[4]

Within just a few years of its discovery, we knew that radioactivity was the source of huge amounts of heat, and we also knew that there was a lot of it on (and in) the earth. This new source of heat invalidated Kelvin's calculations based on a cooling earth. Because Kelvin had assumed that all of the earth's heat came only from the original molten state of the earth or from the sun, he concluded that the heat which was known to be escaping from the earth's surface could only signify that the earth had been warmer in the past and was doomed to grow ever colder in the future. The discovery of radioactivity revealed that the earth's heat was not just residue from its molten origin, and that the lost heat was actually being perpetually replenished by new heat generated through radioactive processes. Even without any quantitation, just the fact of radioactivity's widespread existence in the earth demolished all of Kelvin's chronologies based on the earth's cooling. Radioactivity proved that the earth *wasn't* cooling, and thus opened out its time once and for all.

In 1904, Rutherford gave a talk on radioactivity and its generation of heat at the Royal Institution in London. As he began speaking, he noticed that the eighty-year-old Lord Kelvin was in the audience. Here's how he described it:

> ... To my relief, Kelvin fell fast asleep, but as I came to the important point, I saw the old bird sit up, open an eye and cock a baleful glance at me! Then a sudden inspiration came, and I said Lord Kelvin had limited the age of the earth *provided no new source of heat was discovered.* That prophetic utterance refers to what we are now considering tonight, radium! Behold, the old boy beamed upon me.[5]

Not everyone immediately embraced radioactivity's temporal implications, but in the end, they could not be denied. Not only did the discovery of radioactivity vindicate Darwin and Lyell, but it also finally resolved the debate between long and short chronologies for the earth which had begun a century earlier with Hutton and the catastrophists. Once again, it was *heat* that made the case for a longer chronology—this time definitively. Recall James Hutton's concept of repair a century earlier: heat from the earth's interior built the eroded earth back up again through volcanic activity and mountain-building. Now the heat of radioactivity likewise stayed a downward course for the earth—this time from cooling rather than erosion.

By following an odd idea inspired by Roentgen's chance discovery of x-rays, Becquerel—a million miles away from any geological concerns—had done an experiment on luminescent chemicals that within a handful of years had led to the resolution of a controversy about the age of the earth which had endured for over a century. Life does not get much stranger than this . . . except, as we will see in the following pages, it does get stranger. Much stranger.

The first punch delivered by radioactivity had demolished not only a short age for the earth, but the quantitative method itself. We were back to vast eons of time, now no longer just an intuition, but a known reality. Yet those eons remained uncounted. It was now time for radioactivity to deliver its second punch, which would be the coup de grâce and would follow rapidly upon the first. Radioactivity at once demolished the previous quantitative methods and set up in their place a new quantitative method so powerful and accurate that it finally made the measurement of the eons possible. Just as the stable cesium-133 atom serves as an exquisitely precise counter of our daily time, so the unstable radioactive atoms would soon reveal that they also harbor within themselves a constant rate of change that would ultimately count the earth's whole lifetime.

The clock that tells us the time of day and the "clocks" that count the eons share the same principle of counting time by measuring off identical amounts of it. The modern version of our "daily" clock has some other points of connection with the radioactive clocks. Recall Pierre Curie's discovery of piezoelectricity: certain crystals can be made to vibrate at a precise frequency when pressure or an electrical current is applied to them. It is because of this discovery that quartz crystals ultimately wound up as the regulators of the wrist watches we wear today. As a crucial part of an ingenious apparatus that Marie and Pierre rigged up to quantitate the radioactive emissions of the substances they were trying to isolate, that same piezoelectricity also played a central role in the discovery of the nature of radioactivity. Without this measuring device, Marie could never have known that infinitesimal amounts of the then unknown elements polonium and radium also lay buried within the mountain of pitchblende ore from which she had extracted uranium.

Uranium, named after the then newly-discovered planet Uranus, had been discovered and isolated by Martin Heinrich Klaproth in 1789, so the radioactivity of pure uranium could be measured with the apparatus. When Marie put pitchblende into the measuring device, she found that its radioactivity caused a much stronger current than did that of pure uranium, which could only mean that the pitchblende contained at least one other element more radioactive than uranium. As she went about isolating what turned out to be *two* new elements from the pitchblende, Marie measured the current their radioactivity produced in the device, and found that polonium had four hundred times the radioactivity of uranium, and radium nine hundred times.[6] Because of the precision of the piezoelectric measuring device, the amount of radioactivity produced by a substance became a marker which could signal the existence of as yet unknown radioactive elements.

If piezoelectricity represents an interesting mechanical link between our way of telling the time of day and one of the crucial steps on the path toward measuring the eons, a far more fundamental nexus is represented by the *atom*. Although the existence of radioactivity could now be determined, and the amount quantitated, we must remember that at this time almost nothing else was known about it. What caused it? What *was* it? With so little known, it was not unreasonable to imagine that an element's radioactivity might change when it bonded with other elements to form molecules or when its own chemical state changed. Yet as early as 1898, Marie Curie had the correct hunch that radioactivity was a function of matter itself, and not of chemistry. When she tested uranium compounds, she found that the amount of radioactivity was unrelated to the physical or chemical state the uranium was in. It was determined solely by the *quantity* of uranium present, in whatever compound or state. This could only mean that radioactivity was an *atomic* property, and the clock that it would eventually produce would be another form of *atomic* clock.

When we call to get the correct time of day, it is the flipping of the outer electron in the stable cesium-133 atom that is ultimately responsible for its accuracy, and when we open a book on dinosaurs and see the by now familiar period of their reign (from around 200–65 million years ago) it is the regular disintegrations in the *nuclei* of unstable atoms—radioactive clocks—which have brought us this awesome knowledge.

Our story may seem to have a long way to go yet to reach such knowledge, but we are closer than you think. The cesium atomic clock which tells us the time of day arrived long after the basic geography of the atom had been worked out, but the discovery of the radioactive clock went hand-in-hand with the discovery of the nature of the atom itself. It is to radioactivity that we owe the discovery of the neutron, as well as the image of the atom as a kind of miniature solar system—with electrons orbiting a nucleus—which we still hold today. When Henri Becquerel discovered radioactivity in 1896, the atom was still believed to be a single, indivisible entity. It was not until the following year that the English physicist J.J. Thomson (1856–1940) discovered the first subatomic particle: the electron. Throughout the early period of exploration of radioactivity's nature and mechanisms, the model of the atom was still J.J. Thomson's image of an English plum pudding, in which negatively charged electrons were embedded in a pudding of positive charges.

The vast majority of atoms are stable and virtually perpetual. Every carbon atom you have in your body right now was formed by the collision of three helium nuclei in a star billions of years ago. When the star later became a supernova and exploded, the carbon atoms were thrown out into the atmosphere to ultimately find their way into our bodies. Unstable radioactive atoms make up only a tiny percentage of all atoms, and most of them belong to the heaviest elements. It took the enormous energy released by supernovae explosions to forge these heavy elements, but they too were then spewed out into the void, eventually becoming incorporated into the rocks and minerals of our planet. Although there are over a dozen heavier man-made elements, nature herself stopped making atoms with uranium (number 92 of the periodic table), and all of the elements beyond bismuth (number 83) are radioactive. There seems to be an inherent instability in such heavy elements.[7]

In saying this, we are already assuming knowledge that was just coming into being at the turn of the century. In fact, we are able to speak so casually of instability in part because Ernest Rutherford put forth a third hypothesis for what was causing the radioactive rays. Instead of the ideas that energy was somehow being created or that it was being captured from outside the atom, Rutherford suggested that radioactive atoms were changing *inside*—and thus at the subatomic level—and the rays represented the energy being released in the process. In 1902, just six years after Becquerel's chance discovery of radioactive rays, Rutherford had arrived at the correct explanation for them. The theme of *change* which we have followed in the geological and cosmological discoveries from the late eighteenth century onward had now found its way into the interstices of the atom.

Already in 1899, Rutherford published a paper suggesting that radioactive rays were of two kinds, which he called alpha and beta; during the following year he discovered the third form of radioactive emission, the electromagnetic gamma rays. That same year, Marie and Pierre Curie, as well as Henri Becquerel, determined that the beta rays were none other than Thomson's electrons, but it would take another five years for Rutherford to discover that alpha rays were helium (really, helium *nuclei* consisting of two protons and two neutrons which quickly pick up two electrons to become an atom of helium).

Although the emission of these rays certainly signaled change, Rutherford's transmutation theory was based on something more. Both he and another English scientist, Sir William Crookes (1832–1919), had noticed that a second radioactive substance, with different properties from the original radioactive material, seemed to build up in the closed chambers in which they tested the ionization produced by the original substances. In addition to the radioactive element thorium, Rutherford found something else which he called thorium X; in addition to uranium, Crookes found uranium X. The Curies still resisted Rutherford's "alchemical" idea of transmutation, but by 1903 the preponderance of the evidence pointed in its direction. Base metal wasn't being transformed into gold according to the ancient alchemical dream, but one radioactive element was indeed being transformed into another.

Today we know the explanation: an element is defined by the number of protons in its nucleus, so that when it loses an alpha particle (two protons and two neutrons), it necessarily becomes another element. When it ejects an alpha particle, a uranium atom might be said to commit suicide, because from that moment it literally ceases to exist, and in its place is born a thorium atom, which has two fewer protons in its nucleus. Similarly, in beta decay, one of the (neutral) neutrons in the nucleus emits an electron and thereby becomes a positively charged proton. Thus, its nucleus now has one *more* proton than it did before, and has literally become another element. These processes, known as radioactive decay, make sense of the mysterious thorium X and uranium X: they were in fact new elements which had been created by the loss of alpha or beta particles from the nuclei of the original materials. Thus the subatomic change in radioactive decay involves only the atom's *nucleus*.

Marie Curie's ability to quantitate the radioactivity coming from a source allowed her to confirm the existence of unknown radioactive elements, but she, and others, soon realized that radioactivity had an even more precise signature. Already in 1902, when a German scientist named Willy Marckwald decided he had discovered another radioactive element, Marie was able to disprove it by showing that his "new" element had precisely the same 140-day half-life as the polonium she had discovered several years earlier.[8]

Early on, it became clear that the decay of a radioactive substance occurs at a very constant rate, and in this sense a radioactive clock is exactly like the clock that tells us the time of day. Just as we know that every hour the traditional clock will tick off sixty equal minutes, so also we know that in any sample of uranium about 1.5 percent of its uranium-238 will decay to lead-206 every 100 million years.[9] Another way to express this would be to say that uranium-238 has a half-life of 4.5 billion years.

Unlike the linear rate at which the early candle clocks melted or water clocks drained or our wrist watches tick today, the constant decay rate of radioactivity is exponential. A radioactive substance's half-life is simply the amount of time required for half of its atoms to decay into another substance. The same amount of time is required for the remaining half to lose half of its atoms, and for half of that half (one quarter of the original) to lose half of its atoms, and on and on. Although theoretically there will always be some of the original material left, after about ten to fifteen half-lives there usually isn't enough remaining to count.

Since each atom of the original material (called the "parent") becomes a single atom of the "daughter" substance, the total amount of atoms remains the same. At the same constant rate at which the parent loses atoms due to the emission of alpha or beta particles, the amount of the daughter substance increases. Although in reality you couldn't find so small an amount, imagine finding a rock containing a million atoms of a particular radioactive isotope with a half-life of a billion years. After a billion years, there will be 500,000 parent atoms and 500,000 daughter atoms. In another billion years, the remaining 500,000 of the parent will again divide so that there will now be 250,000 of the parent and 750,000 of the daughter. When the parent again halves after another billion years, only 125,000 atoms of the parent will remain, but the daughter atoms will have grown to 875,000. However many half-lives are gone through, the total number of atoms will always remain one million. The geologist, having already calculated the billion-year half-life for the parent substance in the laboratory, can then analyze the rock, find the 125,000 parent and 875,000 daughter atoms, and then use a simple formula to determine that the rock has undergone three half-lives and is thus three billion years old.

The exponential decay rate makes sense when you realize that the more atoms there are, the more decays there will be, but the *percentage* of the total number of atoms which decay will always remain the same. (In terms of half-life, that percentage is fifty percent.) Thus, the same amount of time is required for half of 250,000 atoms to decay as for half of a million atoms.

Although there may be a whole chain of intermediate decay products, the parent–daughter pairs referred to in radiometric dating are the original radioactive parent and the ultimate stable daughter in which the decay series culminates. Many such parent–daughter pairs have been discovered, and each constitutes an independent clock which can tell us the age of the rock in which they have been sequestered for eons.

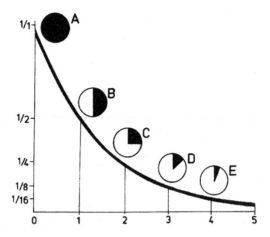

30. Every 1,310 million years, half of the atoms of potassium-40 decay to argon-40, so that the ratio of parent–to–daughter atoms in a sample reveals its age. A. Original sample. B. After 1,310 million years, half of the potassium-40 atoms remain. C. After 2,620 million years (two half-lives), one-quarter remains. D. After 3,930 million years (three half-lives), one-eighth remains. E. After 5,240 million years (four half-lives), only one-sixteenth remains.

In the case of the potassium–argon clock, there are no intermediate products—radioactive potassium-40 changes directly into the stable gas, argon-40. On the other hand, an atom of uranium-238 is transmuted into *thirteen* other radioactive isotopes before finally becoming a stable atom of lead-206, the ultimate destination of the whole cascade toward stability which begins with uranium-238's emission of an alpha particle. Although in practice there are at least as many obstacles to overcome with these clocks as we found with the traditional clock, the basic principle of using radioactive decay to measure time is as simple as this.

Clearly, a radioactive clock is a *statistical* phenomenon. No one will ever be able to predict when a *particular* atom will decay, but there are trillions of atoms in even the tiniest amount of any substance. The probability that each atom will decay in a certain period of time is constant, so that the half-life for the vast aggregation of atoms that *is* the substance always remains constant. Insurance companies rely on the same principle. Although they can't know which of their individual policy holders will die in a particular month, statistics tell them that they will have to pay off a certain number of beneficiaries every month.

It is the huge number of atoms in any substance that allows half-life calculations to be made. Consider that a few grams of an element (and there are 28.3 grams in an ounce) contain about a septillion atoms: 1×10^{24} or 1,000,000,000,-000,000,000,000,000.[10] Today, with these huge numbers and an electronic measuring device to either count the emissions of alpha or beta particles over a period of time or, alternatively, to count all of the daughter atoms created, one can as easily calculate uranium's 4.5 billion year half-life, or even rubidium-87's half-life of 48.8 billion years, as polonium's of 140 days.

The calculation of half-lives of pure materials in the laboratory is actually the "easy" part. The difficulty comes with finding in nature parent–daughter pairs which form a closed system, i.e., one which has remained insulated from disruption by (or leakage to) the outside for eons. To give just one example: in one of the most widely used clocks—the potassium–argon pair—the argon daughter is a gas. Some of it may leak out of the system if the rock is heated, and thus the rock will seem younger than it actually is. Recall the tremendous efforts to isolate the pendulum from outside forces—most of the difficulties of the radioactive clocks are based on the same problem.

The invariant signature half lives of isotopes range from millionths of a second to minutes to hours to days to years to thousands to millions to billions of years. To give a few examples: beryllium-8 has a half-life of 0.0000000000000002 seconds, that of aluminum-26 is 7 seconds, tritium's is 12.5 years, while carbon-14's half-life climbs to 5,730 years. At the other end of the scale, potassium-40's half-life is 1.31 billion years. To show the difference that three neutrons can make: uranium-235's half-life is "only" 704 million years, as compared to uranium-238's of 4.5 billion years. Also recall that rubidium-87 has a half-life of 48.8 billion years, while that of rubidium-84 climbs to 51 billion. One of the Methuselahs is samarium, with a half-life of 106 billion years. We don't yet understand *why* a particular isotope has the half-life it does, but we do know that it is constant in a way that the accumulation of salt in the ocean or the rate of sedimentation can never be.

We have begun using the word *isotope,* so let's explain that isotopes—discovered by Frederick Soddy in 1913—are simply atoms of the same element having different numbers of neutrons in their nuclei and therefore slightly different masses. Because they have the same number of protons, they are the same element, and thus have identical chemical behaviors, but they may have very different radioactive properties because of the differences in their nuclei. It is not only the heavy radioactive elements whose isotopes are radioactive—many predominantly stable elements may have an isotope that is radioactive. In fact, two of the most important isotopes used in radiometric dating—potassium-40 and carbon-14—make up a tiny percentage of the atoms of these otherwise stable elements. The number after the isotope's name indicates the total number of protons and neutrons in its nucleus.

A desire to verify the apparent constancy of the rate of radioactive decay arose early. In 1907, Rutherford put some radon into a bomb to test the steadiness of its decay rate. The bomb generated enormous pressure and a temperature of 2,500° C when it exploded, but the radon's decay rate remained unchanged. In Holland, H.M. Kamerlingh Onnes (1853–1926) did studies on radioactive substances at very low temperatures. Marie Curie traveled to Leiden to work with him. In 1913, they published a paper showing that radium's decay rate at the boiling point of liquid hydrogen (–252.8° C) did not change by more than 0.05 percent, if at all. There were also experiments to test the effect of gravity: radioactive samples were put on mountain tops or deep in mines, and were whirled in centrifuges. Other samples were subjected to powerful magnetic fields. In every case, the radioactivity of the substances remained unchanged.[11] Speaking of these tests and earlier studies on the effect of chemical changes, Arthur Holmes (1890–1965) had this to say:

". . . the [radioactive] atom continues to obey the law which determines its life period, unmoved by any experiences through which we can oblige it to pass. The radioactive properties are simply superimposed upon the chemical and other properties of the substance. . . ."[12]

Just as advertisers today use the gambit of putting their watches through every form of torture test, so radioactivity's early students imposed extreme conditions upon it to try to dislodge its steady rate of disintegration. At the end of all their tests, they realized they had discovered one of the very few invariant phenomena of nature. (The previously mentioned leakage of argon gas and other disruptions to a closed parent–daughter system affects only the *quantity* of atoms, not their decay rate.)

Today we understand *why* the radioactive clocks tick on with a constancy impervious to any pressures from outside. Radioactivity involves only the *nucleus* of the atom, and that nucleus is insulated from the outside world by the cloud of electrons which surrounds it. Ordinary chemical reactions of the atom involve only the outermost electrons, which is why its radioactivity is unaffected when a substance combines with another to form a compound.

In addition to the protective layer formed by the electrons, the nucleus is further shielded from outside forces by the enormous energies required for nuclear change to occur. The force required to affect a nucleus is a *million* times greater than that required for the chemical bonding of atoms to form molecules. Were it any easier for humankind to force change in nuclei, powerful nuclear reactors and particle accelerators would not be needed. Such energies are far beyond those involved in the processes of nature's formation and degradation of the earth's rocks.[13] Thus, the radioactive clocks have been able to tick off their constant half-lives, immune to the trauma of their passage from the depths of oceans to the mountains' heights, through boiling heat and stunning cold, through enormous pressures and changing magnetic fields, for millennia upon millennia.

In 1905, less than a decade after Becquerel discovered radioactivity, Ernest Rutherford and Pierre Curie both suggested that its constant decay rate might be used as a clock to determine the age of rocks. As we will see in the next chapter, the idea was surprisingly slow to take hold, but as the century wore on, and more and more clocks correlated well with each other and with the relative ages of the geological strata, the quantitative knowledge of the earth's history which we have today gradually came into view. The geologists had pointed the way, but in the end it turned out to be the physicists—*nuclear* physicists—who found the answer to the age of the earth. The immense amount of time Darwin knew was required for the evolution of the fossilized species found in the geological column would turn out to be only about an *eighth* of all the earth's time.

The clock which tells us the time of day is a relative phenomenon. Zone time, which had its birth not long before the discovery of radioactivity, came into being because people agreed to see time in new terms. Although it certainly incorporated a new awareness of the relationship of our turning planet to time, this new everyday time was as much a matter of utility as of knowledge. The day of a turn-of-the-century man with a pocket watch was no *truer* than the day of an Egyptian telling time by a stick in the ground four thousand years earlier; they just used different languages to describe it. The radioactive clocks involve no such relativism—they would give us *answers* instead of merely find a more efficacious language into which to translate a known phenomenon. Barely two hundred years earlier our planet had been viewed as essentially static, created all-of-a-piece within a handful of days only a few thousand years ago, and worthy of admiration but not of study. The new image of the earth given us by the radioactive clocks would not merely be different from the previous one, it would be far *truer*.

Had Napoleon's army not stumbled onto the Rosetta stone, and had Jean Francois Champollion not subsequently deciphered it, would we today be able to read the ancient Egyptian hieroglyphics? Probably not. The natural clocks created by the constant decay rate of unstable radioactive atoms are another such Rosetta stone. Although there are other reliable short-term methods, *no other means has ever been found to accurately measure the earth's eons.* Who could ever have guessed that this miraculous compact between the infinitesimally small and the infinitesimally large would grow from the fogging of a photographic plate with some uranium salts on it left in a drawer?

Chapter XXI

The Past Recaptured

*That then this Beginning was, is a
matter of faith, and so infallible.
When it was, is a matter of Reason,
and therefore various and perplex'd.*
John Donne (1573–1631)

*To see a World in a Grain of Sand. . . .
And Eternity in an hour.*
William Blake (1757–1827)

Three hundred years intervened between Bishop Ussher's assertion in 1650 that the world was created in 4004 B.C. and Clair C. Patterson's use of meteorites in 1956 to arrive at an age for the earth of 4.55 ± 0.07 billion years. Patterson's figure—still valid today—gives the earth a lifetime over 750,000 times longer than Ussher's. This enormous difference tells us something about the disparate courses of growth in our knowledge of the earth's space and of its time. Columbus discovered the Western Hemisphere in 1492, and by 1522, well over a century before Ussher's pronouncement, Magellan had already circumnavigated the globe. John Harrison's watch found the longitude at sea in 1762, and within another ten years Captain Cook was using that watch to precisely chart the islands in the south Pacific. So carefully were the twists and turns of our planet's surface mapped during the next two centuries, that when the first cameras took pictures of it from space, there were virtually no surprises. It looked just the way the mapmakers said it would.[1]

Our knowledge of the earth's time lagged far behind, and the reasons are not far to seek. In the beginning, the biblical account held such sway over people's minds that they couldn't imagine anything else. No such powerful dogma had stood in the way of the exploration of our spatial world. Later, there was the simple fact that the discovery of radioactivity and its steady rate of decay required a greater scientific sophistication and a far deeper look into nature than did figuring out the spatial dimensions of the planet and then creating a mechanical watch capable of withstanding the rigors of sea travel. On a more basic level, the earth's space is not only visible, but it belongs to the present. It thus makes a more urgent cry for exploration than does invisible time, forever receding into the past.

If we try to compare progress in our knowledge of the two realms of space and time, we might make a rough analogy between the fossil-bearing strata of the geological column (and their relative ages as pieced together by the nineteenth century geologists) and the known spatial world before Columbus made his discovery. Just as people were aware that there was more world to be discovered, so too geologists knew of rocks which appeared to be older than those with fossils, but that is where their knowledge stopped. In reality, the leap which would follow the discovery of radioactivity was far greater than that created by Columbus, since the fossil-bearing strata of the geological column comprise only about thirteen percent of the whole of our planet's time, while over half of the world's landmass was already known before Columbus set out.

We might then suggest that the twentieth century has done for time what the sixteenth century did for space, meaning that in both cases the crucial discovery was made during the last decade of the preceding century, and explorations exploiting that discovery filled the following century. The flaw in this analogy is that although the sixteenth century explorers certainly made maps, those maps lacked the quantitative precision which would only arrive two centuries later, when John Harrison's watch finally allowed longitude, as well as latitude, to be pinpointed.

Radioactivity actually accomplished for time in one step what had required a two-step process for space. Within half a century, not only did the total time of the world become known, but it became known in a precisely quantitated fashion. Our knowledge of the earth's time, which had so long lagged behind that of its space, finally caught up once the Rosetta stone was found which could unlock its secrets. Although our map of time is still rather crude, with many details yet to be filled in, the important milestones are in place.

It was a ticking clock, measuring off equal portions of time, which finally allowed us to measure the space of our planet and today allows us to locate ourselves in outer space. That same ticking of a clock held the answer for time itself. From here on, our story will no longer involve myths, or analogies, or beliefs, or even scientific intuition. It will just be a matter of *counting*. Gone is the ancient analogy of time to the circular motions of the heavens and gone is the rich panoply of Great Cycles which grew from it. Gone as well is the Bible's great story of Creation. The knowledge gained from simple counting has meant the loss—at least in terms of belief—of great works of the human imagination, but in their stead we have gained a knowledge that our imaginations must stretch even to dimly comprehend, much less invent.

Just like the early verge-and-foliot clocks of the fourteenth century, the first radioactive clocks were none too accurate. The problems with the early clocks were two-fold: first, the measuring instruments were still very crude, so that decay rates and half-lives were often miscalculated. For its time, the piezoelectric measuring device was remarkable, but it bears no comparison to today's mass spectrometers which can distinguish isotopes on the basis of the difference in mass made by a couple of neutrons. (Nature's clocks are perfect. What took time was learning how to make sufficiently precise measurements so that they could be correctly "read.") The second problem for the earliest clocks was lack of knowledge. Isotopes were as yet undreamt of, the long series of intermediate steps between radioactive uranium and its ultimate stable daughter lead had not yet been deciphered, and awareness was still dim of the many conditions in nature which can reset the clocks or otherwise cause their readings to be misleading.

Nevertheless, in 1905, just eight years after the phenomenon itself was discovered, the use of radioactivity for measuring time had its birth. Sir William Ramsey (1852–1916) and Frederick Soddy (1877–1956) had just determined the half-life of radium, and already in 1903 it was known that radium emits alpha rays and thus produces helium. Rutherford proposed the first radioactive clock when he suggested that the age of a mineral containing radium might be learned by measuring the amount of helium which had been produced. If no helium had leaked from the system, then by measuring the mineral's present ratio of helium to radium, its age could be calculated.

Using the helium method, Rutherford calculated an age of 497 million years for a fergusonite mineral, and in the same year another pioneer of radiometric dating, R.J. Strutt (1842–1919) calculated a minimum age of 2.4 *billion* years for a thorionite mineral from Ceylon. (The atoms in rocks are bound into natural crystals called minerals, and it is within the crystal lattices of these minerals that most daughter substances become trapped. Rocks are simply aggregates of minerals.) Rutherford and Strutt were already aware that helium can rather easily leak from the crystal lattice, and today this insolvable problem makes the helium method essentially useless. Nevertheless, this clock gave "the first quantitative indications, based on physical principles rather than scientific intuition, that the earth might be billions, rather than tens or hundreds of millions, of years old."[2]

At least as important as this first method was the series of lectures on radioactivity which Rutherford gave at Yale University that same year, because in the audience was an American chemist named Bertram Boltwood (1870–1927). It had already been noticed that lead was always found with uranium, and Rutherford speculated that lead might be the ultimate stable daughter of the uranium decay series. If it was, then another clock would be available, and this one, based on lead-to-uranium ratios, should be more reliable because lead is a solid and thus cannot escape from the system the way helium gas does. Rutherford's idea inspired Boltwood to think about the relationships among the elements which are produced following uranium's decay, and he set out to test the hypothesis that lead was the ultimate decay product of the series. Boltwood found that minerals from the same rock layer (and thus presumably of the same age) always had virtually the same ratio of lead to uranium, and he also saw that the amount of lead increased when he tested minerals from older rock layers and decreased in those from younger layers. The older the rock, the higher will be the daughter–parent ratio because there has been more time for the daughter substance to build up. It appeared that a new clock had indeed been discovered.

Boltwood used the new uranium-lead clock on forty-three minerals and calculated ages from 410 million to 2.2 billion years. Neither Rutherford, Strutt, nor Boltwood claimed precise accuracy for their clocks. However, they knew these clocks were based on a natural phenomenon whose rate of change had a proven constancy which could not be claimed for the nineteenth century clocks based on rates of sedimentation or increasing ocean salinity or cooling of the earth. A modern geochronologist measuring millionths (and even billionths) of a gram of mineral to the third decimal place, and sorting isotopes according to their number of neutrons, would regard these first clocks as crude indeed. He would also acknowledge them as the stepping stones without which his own state-of-the-art chronometers could not exist.

Back in the first decade of the twentieth century, the conceptual leap demanded by these new ages was huge, and few geologists were prepared to take it. Ironically, although they had chafed at the constraint of the figures of 100 million years or less dictated by Kelvin and the other nineteenth century "clockmakers," they had gotten used to it and now had trouble accepting this new plenitude of time. There was also the human problem of lifetimes of work going down the tubes. The scepticism and outright indifference with which the first radioactive clocks were greeted was thus in stark contrast to the almost universal enthusiasm for the first mechanical clocks to tell the time of day. Even the "inventors" themselves showed only lukewarm interest. Rutherford published about one paper per decade on the subject, and Boltwood left off work on his lead clock and returned to his efforts to unravel the other steps in the decay series. Strutt worked on fine-tuning the helium method until 1910, but then he, too, left the field of radiometric dating.[3]

If Strutt himself was not to continue the work, he left it a great legacy in the person of his pupil Arthur Holmes, who almost single-handedly kept interest in radiometric dating alive during the next couple of decades. In 1913, at the age of 23, Holmes wrote an extraordinary book entitled *The Age of the Earth*, in which he made the case for the radioactive clocks over all previous dating methods. He would devote his whole life to refining the clocks and to pressing for acceptance of both their viability and validity as a means of discovering the earth's history.

To appreciate another of Holmes' early contributions, we must return for a moment to the first half of the nineteenth century. Earlier we spoke of the great insight of William Smith and Georges Cuvier that sedimentary strata could be put into temporal order on the basis of their fossil content, but we didn't say much about the arduous century-long effort by other geologists throughout Europe to identify and order the major layers of strata and give them the names we still use today. The job was mostly completed by mid-century, but in 1879, when the Ordovician system was finally put into place, the geological column with which we are familiar was finally complete. Already in 1841, John Phillips had proposed the larger divisions of the Paleozoic, Mesozoic, and Cenozoic eras, meaning "ancient life," "medieval life," and "recent life." The placement of the boundaries between the divisions was based on the dramatic changes in the fossils resulting from the massive extinctions at the ends of the Permian and Cretaceous periods, and the beginning of the Paleozoic was determined by the abrupt increase in fossils at the start of the Cambrian.[4]

Thus, by around the middle of the nineteenth century, the world-wide history of past (fossilizable) life on earth had already been ordered into a construct called the geological column. (Today we know that single-celled life may have appeared as early as four billion years ago, and soft-bodied multicellular organisms existed by 1.4 billion years ago, but it was only when animals developed hard parts like shells—around 570–550 million years ago—that they could be easily fossilized.) The creators of the geological column knew only that there was a huge amount of time involved. They had no idea *how much*. Imagine how your sense of human history would wither away into murkiness if you only knew that Egypt's great age came sometime before the classical Greek period, which in its turn came sometime before the Roman Empire, which itself was separated from the medieval world by a rather long, blankish period, and on and on to the present. Without the quantitative measure-

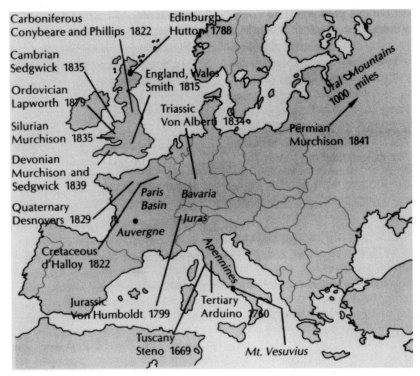

Carboniferous
Conybeare and Phillips 1822

Cambrian
Sedgwick 1835

Ordovician
Lapworth 1879

Silurian
Murchison 1835

Devonian
Murchison and
Sedgwick 1839

Quaternary
Desnoyers 1829

Cretaceous
d'Halloy 1822

Jurassic
Von Humboldt 1799

Edinburgh
Hutton 1788

England, Wales
Smith 1815

Triassic
Von Alberti 1834

Ural Mountains
1000 miles

Permian
Murchison 1841

Paris
Basin

Bavaria

Juras

Auvergne

Apennines

Tertiary
Arduino 1760

Tuscany
Steno 1669

Mt. Vesuvius

31. Lines indicate the place where each geologic system was first identified. The names and dates beneath each system indicate the discoverer and year of discovery. By following the fossil sequence, these strata were eventually put into the chronological order represented by the geological column.

ment of the years, our civil history would all unravel into a vague "long ago." That was just the situation with the geological column. Its ages were only relative. What was needed were the *absolute* ages which only the radioactive clocks could give.

Although Boltwood verified that the ages of his forty-three minerals were consistent with the relative ages of the geological column, he didn't use them to date the strata. In 1911 Arthur Holmes made the first of many such efforts he would make throughout his long career. The dates he arrived at weren't nearly as accurate as the ones we have today, but they were extraordinary given the state of the art of his clocks, and

HELIUM METHOD.   **LEAD METHOD.**

	←Pleistocene.	
Basalt, Oregon. = 13	Pliocene	
Basalt, Oregon. = 18	Miocene	
Basalt, Germany. — 32	Oligocene	Uraninite, Mexico. 34 — Brannerite, Idaho.
	Eocene	
		70 — Pitchblende, Colorado.
Basalt, N. Mexico. — 83	Cretaceous	
Post-Nevadan Dyke. — 96		------100 million years.
Nevadan Granodiorite. — 107	Jurassic	123 — Ishikawaite, Japan.
Basalt, Nova Scotia. — 135		
Dolerite, N.Jersey. = 161	Triassic	
Dolerite, Conn. = 163		
Oldest Basalt, . — 175 N. Jersey.		
200 million years.------	Permian	
Basalt, Mass. —224		220 — Thorite, Norway. Pitchblende, Bohemia.
Basalt, Shropshire. —234		232 — Uraninite, N. Carolina.
Basalt, Shropshire. —256	Carboniferous	
		269 — Pitchblende, Silesia. 278 — Various Minerals, Connecticut.
Volcanic Rock, Mass. —292	Devonian	
		------300 million years.
	Silurian	
Dolerite, { —348		349 — Uraninite, Mass.
Pennsylvania. {	Ordovician	Cyrtolite, New York.
{ —365		366 = Uraninite, 371 Branchville, Conn.
400 million years.------		395 — } Kolm, Sweden. 405 —
Basalt, Virginia. { —427	Cambrian	
{ —453		

32. Arthur Holmes' radioactive dates for the geological column using the helium and lead methods (1937). Far more accurate dates would be found once isotopic clocks began to be used.

the fact that the sedimentary rocks in which fossils are found are the most difficult of all rocks to accurately date.

In 1917, just about the time that the wearing of wrist watches by World War I soldiers had finally persuaded civilian men that they weren't too effeminate, the first really exhaustive radioactive calibration of the fossil record was made by a geology professor at Yale, Joseph Barrell (1869–1919). Barrell collected all the known radiometric ages for rocks and used them to put remarkably accurate dates on the geological column. In the process, he turned the tables on the earlier scheme of gauging time by the rate of sedimentation when he used the radiometric dates to show that average sedimentation rates are far, far slower than previously calculated.[5]

The accuracy of Barrell's geological column was particularly amazing given the problems of dating sedimentary rocks. The radioactive clocks in the mineral grains of igneous rocks are set as soon as the rock cools down enough for the mineral to crystallize. As Arthur Holmes put it: "Radioactive minerals . . . are clocks wound up at the time of their origin."[6] Assuming the rock wasn't reheated to the point where it again became molten and caused its clock to be reset to a new origin, the clock kept on ticking within that same rock until now. But sedimentary rocks are made up of silt which has been washed away from the igneous rocks where it originally formed. A radioactive clock in a mineral grain found in a 100-million-year-old sedimentary layer may have been washed away from a rock that cooled 500 million years ago. That clock will try to tell us that the sedimentary strata in which it is embedded is also 500 million years old. A few minerals, like glauconite, have now been identified which crystallize at the time the sediment is laid down. These can give an age directly from the sediment, but for the most part sedimentary rocks can still only be dated indirectly, by the clocks in lava flows or other igneous intrusions into sedimentary layers. Awareness of the pitfalls in dating sedimentary rocks and the strategies to overcome them is far greater today than it was in the early years of the century.

Our familiar geological column combines the work of two different groups of people, working in two different disciplines, a century apart. It is a whole greater than the sum of its parts, and the loss of either part would destroy its power over our imaginations. Without its radiometric dates, the column would dissolve into a turbid vagueness, but without the fossils and the different strata, the dates would merely be scattered over blankness, an empty form enclosing no content.

That is somewhat the problem with the older rocks. What of the rocks that Strutt and Boltwood found to be more than two *billion* years old? These rocks have virtually no fossils and even today, when far older rocks have been found on all of the continents, this period before animals with fossilizable hard parts—which makes up about eighty-seven percent of the earth's history—feels to us a bit like a no-man's land in comparison to the thirteen percent which reveals the vibrant pageant of the evolution of life. Nevertheless, it would be these older rocks which would lead us to the age of the earth. *The geological column is no help whatsoever when it comes to the question of the earth's age.* If the great contribution of the nineteenth century was the piecing together of the geological column, the twentieth century, with the crucial help of the radioactive clocks, would put the temporal domain encompassed by the column into perspective by bringing into view the eighty-seven percent of the earth's lifetime which preceded it. During the twentieth century our ideas have changed most dramatically about this earlier, Precambrian pe-

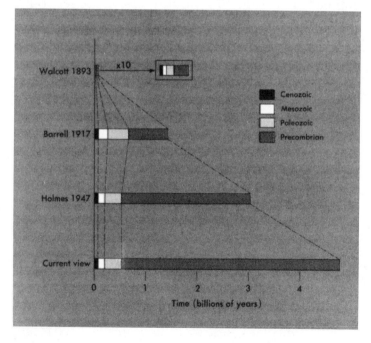

33. Changing conceptions of the magnitude of Precambrian time in the twentieth century.

riod. As the century has worn on, we have revised and refined the dates on the geological column—a few million years here and a few million years there—but the period *before* the fossil record has tripled since Barrell's calculation in 1917.

Meanwhile, the scepticism about radiometric dating continued. Holmes, Barrell, and a handful of others kept it afloat. The tide finally began to turn only after 1931, when the National Research Council of the United States National Academy of Sciences appointed a committee to review the issue of the radioactive clocks. The committee's report finally declared their unanimous agreement that the clocks offer the only accurate and reliable timescale for the earth's history.

Before we turn to the solution to the age of the earth itself, we must look briefly at some of the discoveries around the middle of the twentieth century that helped to make it possible. The basic principle of the mass spectrometer was understood at the turn of the century, and instruments were built throughout the following decades, but it was the high resolution machine developed by Alfred Nier in the 1930s and 1940s which finally made an impact on radiometric dating. The mass spectrometer sorts charged particles according to minute differences in their masses by sending them through a magnetic field—the lighter the particle, the more it will be deflected by the field. Because it can separate atoms of the same element on the basis of their number of neutrons, the mass spectrometer was able to discover new isotopes—both radioactive and stable. Up until now, radioactive clocks had used whole elements, which contained more than one isotope. The next generation of clocks would use only isotopes, and thus would gain a far greater precision.

The mass spectrometer also allowed effective clocks to be made out of tiny amounts of material. Before, a tenth of a gram of parent–daughter substance was needed, but the mass spectrometer can deal with clocks weighing a few millionths of a gram. Because radioactive minerals can often only be found in trace amounts, this allowed far more rocks to be dated than had previously been possible. As if the great ages indicated by the radioactive clocks were not hard enough to fathom, we must also try to grasp that they have been wrested from a dot of substance no thicker than a human hair!

Most of the new clocks were also discovered before mid-century, but their use in dating geological material only got underway during the 1950s. It has continued unabated ever since. Among the new radioactive clocks are potassium-40–argon-40 (half-life of 1.3 billion years), rubidium-87–strontium-87 (48.8 billion-year half-life), samarium-147–neodymium-145 (10.6 billion years), and rhenium-187–osmium-187 (45.6 billion years). The elements involved run the gamet: although radioactive potassium-40 makes up only 0.01 percent of all its atoms, potassium is one of the most common elements in the earth's crust, while samarium and rhenium are among the rarest. For dating purposes, it doesn't matter. All that matters is that the parent isotopes are radioactive, that they have long half-lives (so that some of the parent remains to be counted after so many millennia), and that enough substance can be found for both parent and daughter isotopes to be measured.

These new clocks gave another boost to the number of rocks which can be dated, because not every rock contains uranium. At least as important was their ability to validate the age registered by another clock in a rock in which two or more clocks can be found. Just as in daily life we call to get the correct time by the cesium atomic clock, or ask another person what his watch says in order to verify our own, so too is cross-checking important in radioactive time-telling. There are many rocks in which two or more of these clocks can be found, so that if they both show the same age for the rock, geochronologists can be sure their measurements are accurate and that both parent–daughter systems have remained closed to the outside.

It is this need for cross-checking that has caused the "old" uranium-lead clock to remain the preeminent radioactive time-counter. The mass spectrometer revealed that uranium harbors within itself not one, but *two*, radioactive clocks. We already know that half of uranium-238 decays to lead-206 in 4.5 billion years, but uranium-238 makes up only 99.27 percent of natural uranium. The remaining 0.73 percent is uranium-235, half of which decays to lead-207 in 713 million years. You will never find uranium without also finding both clocks, so that uranium—alone among all the radioactive time-keepers—has a *built-in* cross-checking mechanism to prove or disprove the validity of a measured age. Because their half-lives are so different, if there has been any leakage of daughter isotopes from the system, the ages of the rock calculated by the two different sets of parent–daughter isotopes will differ. If they are the same, we can be certain that the system has remained closed and the age told by both clocks is valid.

Were uranium an extremely rare element in the earth's crust, this gracious gift of nature would not be of much help. But uranium is actually quite widespread, and most important, it is found in the mineral zircon. In addition to containing uranium, one of the several other beauties of zircon for radiometric dating is that it melts only at the extremely high temperature of 750° C, a temperature which the earth has rarely reached even in its most tumultuous periods. Most other minerals melt at far lower temperatures, and their clocks have thus been reset as many times as they have melted. Such resettings of the clocks, which are readily apparent to geochronologists, are important sources of information about events in the earth's history, but they are less helpful when greatest ages are wanted. Tiny grains of impervious zircon can remain closed systems throughout eons.

In recent years, more and more sources of zircon have been discovered. This, combined with technology that allows accurate dates to be wrung from tinier and tinier traces of material, now allows many, many igneous rocks to be dated by the uranium-lead clock. Such absolute dates do more than refine the geological column or send up flares into the dark night of the Precambrian period. They have also helped us to learn where our continents have drifted over the eons. At odds with the land surrounding it, a sandstone underlying Nova Scotia shares its radiometric age with sandstones in North Africa, a strong indication that the two were bound together before the opening of the Atlantic Ocean split them apart. Radioactive clocks also tell us that the Appalachian mountains, along with parts of Newfoundland, Ireland, England, and Norway, were once connected within the same geological formation.

Wonderful as this new knowledge is, it resembles the geological column in being knowledge of the midstream and not of the source. We must now finally let the radioactive clocks take us so deep into the depths of the Precambrian that they will finally find its end-point. There never was and surely never will be any other guide to lead us into this wilderness. Without the clocks, it would forever have remained unknown.

The earth is an active, living planet on which old rocks are forever being destroyed and new ones formed. The young terrains of the Cambrian, which were created in only an eighth of the earth's lifetime, have taken over three-quarters of the surface of today's continents. Nevertheless, outcrops of Precambrian rocks can still be found on all the continents, usually in huge upland areas called shields. We are most familiar with the vast Canadian shield lying below Hudson Bay. Precambrian rocks are also exposed at the bottoms of deep gorges like the Grand Canyon and in various smaller outcrops. It was in such regions that Boltwood, Holmes and the other pioneers found their two-billion-year-old specimens.

The oldest rock found on the earth certainly tells its *minimum* age, and the search for the oldest rock continues to this day. Rocks over 3.5 billion years old have now been found on every continent. In 1971, 3.7 billion-year-old rocks were discovered in western Greenland, but recently the torch has passed to 3.9 billion-year-old rocks in Canada's Northwest Territories. Not only do these rocks tell us something about the age of the earth, but they also give us knowledge about what was happening upon it. When the earth first formed it was molten, so that the existence of these rocks by 3.9 billion years ago proves that some of the earth's molten surface had cooled and hardened into continental crust by that time.

If rocks older than these haven't yet been found, *signs* of older rocks have. Applying Hutton's uniformitarian idea that the processes we see today must also have been going on in the past, geologists reasoned that even the earliest continents must have had beaches, and they searched ancient sediments in the hope of finding mineral grains washed into them from even older rocks. The search paid off in western Australia, where they found a 3.6 billion-year-old sandstone containing grains of zircon ranging in age from 4.1 to nearly 4.3 billion years. Since zircons are common in continental rocks like granite but almost nonexistent in the basalts of sea floors, we can infer that some form of continent existed almost 4.3 billion years ago.[7]

The search for the oldest rock could go on forever, but the only answer it will ever be able to give us is the *minimum* age of the earth. We know that the newly-formed earth was highly volatile—not only was it very hot, but it was continually being bombarded by nebular debris which its gravitational force pulled in from space. It's highly unlikely that material from this earliest period could still exist.

Were we doomed to eternal ignorance about the time of our planet's birth because of this problem? We weren't, but rescue came only when we turned our attention away from the earth and toward the sky. Like fossils, meteorites had been observed since antiquity, and had also been a source of puzzlement and disbelief. The sky was air, so how could solid stones fall out of it? When there was a meteor shower in New York late in the eighteenth century, the Virginian Thomas Jefferson declared that he could more easily believe that Yankee scientists were liars than that stones could fall from the sky.[8]

Today we know that stones indeed fall from the sky, and we think we know why. Scientists now believe that most meteorites are pieces which have broken off from asteroids, those larger chunks of cosmic debris which circle the sun in a broad belt between the orbits of Mars and Jupiter. At some point, a smashup in the asteroid belt knocked the meteorites out of their orbits and eventually set them on a collision course with the earth. Part of the beauty of these falling stones for radiometric dating is that they have lived their lives in the deepfreeze of outer space, so that their clocks have rarely been reset after melting as have the clocks in the rocks of our volatile planet.

There are several categories of meteorites, the most primitive being the chondrites, which make up eighty percent of all meteorites. Chemical analysis of chondrites has revealed that their composition very closely resembles that of the sun, except for the lower levels of gaseous hydrogen and helium which can't be held in solid rock. Thus, an even greater beauty of the chondrite meteorites is that they seem to be samples of the primordial matter from which the earth and the other planets formed.[9] If we could know their ages, it would surely give us an idea about the age of the earth.

In the mid 1950s, just about the time that the first cesium atomic clock for telling the time of day was built, meteorites were first dated by Clair Patterson of the California Institute of Technology and were found to be 4.55 billion years old. Patterson used a variant of the uranium-lead clock, and since then his figures have been corroborated many times over by several of the other clocks. Today, because meteorites can be seen so easily against its white snow, Antarctica has become their major hunting ground. Speaking recently of the 4.5 billion year age, one of the Antarctic meteorite hunters said: "The uniformity of that number is amazing. They may vary 100 million years here and there, but no chondrites are 3 or 5 billion years old."[10]

Way back in 1956, Patterson wasn't satisfied with simply making the reasonable assumption that the earth must necessarily be as old as the meteorites. He gathered earth rocks which came as near as possible to having the average lead isotope abundance for the earth, and he compared their lead isotope composition with that of the meteorites. He found that the composition of lead isotopes in both the earth and the meteorites was so close that they must have been formed at the same time. This fits the currently accepted hypothesis for the formation of the solar system: all the planets and asteroids formed in a very short period of time after the condensation of silicates out of the nebula surrounding the newly-formed sun.[11]

Although already in the eighteenth century scientific thinking had tended toward the idea that the earth had an origin in time, the possibility that it might be infinite had still not been definitively disproved. The sequential changes in fossils had been a first clue that our time is linear, but like the miles of sedimentary strata in which they are embedded, they could not comment on the possible *beginning* of earth's time. Now, finally, the radioactive clocks proved that the earth's linear time was not infinite when they all stopped at the 4.5-billion-year point. The earth had a birth, and it had at long last been found.

In the half century since Patterson's landmark discovery, all the new evidence has only corroborated it. Consider the moon. Unlike the earth, the moon is not an active planet where the past is constantly being eroded away, and it has been proven that most of the craters we see upon it were formed very early in the moon's history when it was still being heavily bombarded by nebular debris. Thus, the moon should have many rocks far older than can be found on earth, but the early bombardments would have reset some of the clocks which could have registered its birth. The rocks brought back by the Apollo expeditions tell us just that. Many rocks were dated to 3.5–4.0 billion years, and a few in the highlands, which weren't cratered by falling debris, registered dates ranging from 4.42–4.51 billion years according to several different radioactive clocks.[12]

Today, the only viable hypothesis for the moon's formation is that it coalesced from debris kicked off into space when a huge planetoid collided with the earth soon after its formation. If the moon formed *from* the earth and *after* the earth, and it contains rocks as old as 4.5 billion years, then the earth must necessarily be that old, too.

In complete contrast to the geological column, the Precambrian period contains only one universal boundary line which separates the Archean eon from the Proterozoic at 2.5 billion years ago. Each of these vast periods thus makes up about half of the eighty-seven percent of the earth's lifetime which preceded the period encompassed by the geological column. Yet there is nowhere in the world that you can actually lay your hand on this boundary as you can on all of those within the geological column. It is just a convenient number

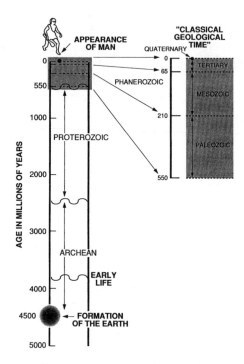

34. The 550–570 million years comprised by the fossil-bearing strata of the geological column make up only about thirteen percent of the earth's whole lifetime.

to indicate some broad changes in chemical composition of the rocks and in the nature of the few fossils which can be found.[13] The radioactive clocks have carried us across this terra incognita to the earth's origins, but it is unlikely that they will ever be able to provide a correlated worldwide story for the Precambrian. Even today, the Precambrian shields on different continents each have only local names for their different temporal epochs. Without a system of signposts like that which fossils provide for the periods of the geological column, our knowledge may be forever limited.

We will never be able to chart the earth's time with the mile-by-mile exactitude that can be found in any modern atlas of its surface, but what our temporal map lacks in fine-tuning, it makes up for in its power to open to our minds' eyes a world that would otherwise have forever remained closed to us (or at least have been revealed only as the highly truncated and murky depths indicated by the undated fossil record). We can go out and verify what our maps of the earth's surface tell us, but the lost worlds conjured by the constant ticking of the radioactive clocks must forever remain a treasure of knowledge which we only need strong minds, and not strong legs, to travel to.

Chapter XXII

Finding the Time of Humankind

Descended from the apes! My dear, let us hope that it is not true, but if it is, let us pray that it will not become generally known.

Wife of the Bishop of Worcester

The eons that the radioactive clocks have captured for us stand before our minds with the same brutal, frightening power we feel when we try to think of the light-years of bitter cold void which separate the violently birthing or dying stars of our own galaxy, and then of the billions of galaxies beyond, with their own. . . . It is all too much.

Compared to the previous four billion years of our earth, we may want to try to make the last half billion or so years of the pageant of life feel almost familiar, almost home ground. But we can't. That would be to use the geological column as we use our wrist watches—to take the sign for the thing itself and forget the awesome reality which it only symbolizes. With continents knocking into one another, with climates and magnetic fields switching back and forth, with the dinosaurs kings of the earth for nearly 150 million years, we can't. Even the very earliest forebears of our own species didn't branch off from the African apes until about five million years ago. That's a measly number compared to the dinosaurs, which themselves are as nothing to the whole reign of life, which in its turn is lost next to the 4,000 million years which came before. No wonder we are hunkered down reading our wrist watches. It is just too much. The eons are our world, but they are beyond our grasp. They are not *us*.

Throughout the nineteenth century, as the pieces of the geological column were slowly falling into place, there was a parallel growth in knowledge of our own "prehistory." Although the word didn't gain common usage until mid-century, the reality which it named had been glimpsed by an insightful few throughout the previous several centuries. Most people, however, looked at the clues and saw only what tradition told them to see. Just as fossils had always been there for the looking, so too had ancient artifacts made of stone or bronze or iron. Unlike wooden implements, these did not decay. When they were found from time to time in the earth, people puzzled over them just as they had puzzled over fossils, and they explained them in the same supernatural terms: they were fairy arrows, elfshots, or possibly thunderbolts. By the seventeenth century, the explanations had gained a "scientific" tinge. One contemporary, Ulysses Aldrovandi, explained stone tools as "due to an admixture of a certain exhalation of thunder and lightening with metallic matter, chiefly in dark clouds, which is coagulated by the circumfused moisture and conglutinated into a mass (like flour with water) and subsequently endurated by heat, like a brick."[1]

As colonization and exploration of America progressed, the clear similarity between the Indians' stone tools and weapons and the artifacts unearthed in Europe strengthened the idea that they might be man-made. It also increased people's curiosity about them, and by the eighteenth century, papers were being published in both France and England that suggested these artifacts were the tools of ancient peoples who had lived before Adam. Some even went so far as to propose that there had been Three Ages of prehistory, and that the stone implements had been used by our earliest ancestors, the bronze by somewhat later peoples, and the iron by their successors.

In 1797, an English country gentleman named John Frere (1740–1807) sent some flint implements to the Secretary of the Society of Antiquaries of London and noted that they had been found twelve feet down in undisturbed strata which also contained bones of extinct animals. His own conclusion was that "the situation in which these weapons were found may tempt us to refer them to a very remote period indeed, even beyond that of the present world . . . "[2] The last words surely refer to the Creation or the Deluge, and reveal that human prehistory, like geological time, would have to break through the time barrier imposed by biblical chronology. For the moment, this was hardly a problem, since although John Frere's letter was printed in the journal *Archaeologia* in 1800, it was ignored and then forgotten for the next sixty years.[3]

Nevertheless, the trickle of interest maintained itself, and by 1819 the first truly public acknowledgement of the reality of prehistory came into existence. It took the form of a museum in Copenhagen in which a whole miscellany of artifacts dug up from the Danish countryside was displayed. At the age of twenty-seven, Christian Jürgensen Thomsen (1788–1865), a Danish businessman with no formal credentials for the job other than his previous organization of a coin collection, was put in charge of organizing this mass of objects, and he separated them into the three categories of stone, bronze, and iron. It was the first time that the hypothesis of the Three Ages of human prehistory had been overtly and concretely linked to the objects.

Many scholars scoffed at these categories, believing instead that stone must always have been used by the poor while the bronze and iron objects had belonged to the well-to-do. Yet when Thomsen's protegé Jens Jacob Worsaae (1821–1885) excavated throughout Denmark, he discovered a vertical relationship of stone, bronze, and iron objects in the sedimentary strata, confirming the theory of the Three Ages. Worsaae's findings, published in 1840, were followed in the next decade by corroborating evidence from excavations at the site of the Swiss lake dwellers: here, too, stone tools were found in the deeper, older sediments, and the implements changed to bronze and then to iron as the strata became younger.

Without being able to apply any absolute dates, Worsaae could still begin to see that the Stone Age had endured for a far longer time than had previously been believed, and that cruder, unpolished stone artifacts had existed for millennia before the polished implements gathered for Thomsen's museum. Within another two decades, this distinction would be codified into the vastly unequal time periods still known to us as the Paleolithic and Neolithic (literally, old stone and new stone).

Solid as this evidence for man's great antiquity seemed, it was far from being embraced. By now, not only scholars but the general public as well could rather easily accept a long and changing lineage for animals, but they had a far harder time accepting such a thing for humankind. The Bible said we had been created in God's own image, and we surely stood at the apex of the scale of life. In short, *we* were different. And just because the biblical chronology appeared not to hold for the earth itself or for lower animals, that didn't mean it wasn't still valid for human beings. Until the late 1850s, it was almost universally assumed that humans were of quite recent origin.[4]

In 1837, Jacques Boucher de Perthes (1788–1868) discovered chipped stone axes in the same strata as extinct animals in the Sommes Valley of northern France, but his efforts to get others to see what was plainly before their eyes fared no better than had John Frere's attempt nearly forty years earlier. For the next twenty years Boucher de Perthes would campaign for acceptance of the great antiquity of his axes, and for all that time he would be mocked by French scholars.

The following year, when the first of the great prehistoric cave art was discovered near Vienne, France, it was attributed to the Celts, an Indo–European people who spread over much of Europe from the second millennium to the first century B.C. Over forty year later, in 1879, the magnificent bison of the Altamira cave in northern Spain were deemed by scholars to be the work of an itinerant contemporary artist. They appeared to be too fresh—and too full of intelligence and beauty—to have been made by prehistoric humans. It would not be until the turn of the twentieth century, and largely because so much other art had been found in so many more caves, that the cave art would at last be accepted as prehistoric.

This conviction of humankind's recent origins stood in Darwin's way as well. His notebooks from the 1830s clearly show that he believed humans to be a product of evolution, and yet when *The Origin of Species* was published in 1859, he spoke only of animals. As to the widely recognized likeness between apes and man, he only commented that "light will be thrown on the origin of man and his history." He quite reasonably feared that a treatment of humans would turn people away from the more general aspects of his evolutionary theory of descent with modification through natural selection.[5]

Nevertheless, all the reluctance to accept a long prehistory for humankind could do nothing to staunch the mounting tide of evidence. Also in 1859, after more than twenty years of being scorned by scholars in his own country, Boucher de Perthes persuaded some English archaeologists to cross the Channel and look at his stone artifacts in situ. Soon afterwards, a paper vindicating him was read before Britain's Royal Society. Entitled "On the Occurrence of Flint Implements, Associated with the Remains of Animals of Extinct Species in Beds of a Late Geological Period . . . ," this paper officially inaugurated the study of prehistoric man. Little could its authors have guessed the depths of time into which the journey they had embarked upon would lead them.

The world might finally have been ready to accept that the human race had been around at least as long as some extinct mammals, but it had a lot more trouble—as Darwin knew it would—with the skull cap and long bones discovered in the Neanderthal gorge in Germany just two years earlier. Ancient people were one thing, but people anatomically different from ourselves—which hinted at our participation in the evolution Darwin had applied to animals—were another and far more disturbing matter entirely. Although a few scientists realized that these strange, thick bones must have belonged to some form of early human, others insisted they were the remains of anything from an idiot, to a diseased person, a Celt, or a Russian Cossack.[6] Both the ancient Celts and the Cossacks had tribal cultures and were considered by Europeans to be rather barbarous and primitive in comparison to themselves. Perhaps beneath the skin they were as well.

Nevertheless, by the 1860s the tide had begun to turn. Since 1700, fossilized human bones had occasionally been found in the same strata as extinct animals, but they had been almost universally ignored. For all their detractors, the Neanderthal bones were seriously studied by a number of scientists and were in fact the first fossils to be considered as evidence of prehistoric humankind's different physiognomy.

Another indication of the acceptance of prehistory by the 1860s can be seen in the broader culture. At the Universal Exposition in London in 1851 there had been no display relating in any way to prehistory. By 1867, the Universal Exhibition in

Paris showed artifacts from Egypt as well as Europe in its "Prehistoric Walks" section, and in the same year the first International Congress on Prehistory was held in Paris. Collecting and research on human prehistory has not stopped since. To name just a handful of the greatest discoveries: from the discovery of the cave art of Lascaux in 1940, to that of Cosquer and of Chauvet in the 1990s, revelations about our recent forebears have never ceased. And from farther back, to name only a few of the most important fossils that mark the trail of our hominid ancestry: Eugene Dubois' discovery of Java man in 1891, the Taung child found by Raymond Dart in 1924, "Lucy," unearthed by Donald Johanson in 1974, and the Turkana boy's discovery by Richard Leakey in 1984. In whatever form and from whatever period, our past keeps coming back to us.

Not only have the discoveries about our own prehistory kept step with those about our planet, but they have transformed our conception of the world just as spectacularly. Who could ever have guessed, when the Neanderthal remains were discovered in 1857, that less than a century and a half later we would be able to chart the timelines and evolutionary relationships among a whole variety of extinct ancestral hominid species?

During the remainder of the nineteenth century, human fossils were more often stumbled upon than deliberately searched for, but as early as 1887 Eugene Dubois (1858–1940) set out for the Dutch East Indies to dig for fossils of early humans, and four years later he succeeded in unearthing Java man, which would turn out to be the first *Homo erectus* specimen ever found. Even then, so much controversy was aroused when Dubois called this ape-man a human ancestor, that he withdrew the bones from examination and didn't write about them until 1922. Public acceptance of humankind's participation in evolution increased throughout the first half of the twentieth century, fossil hunts were organized in South and East Africa as well as in China, and a vast array of hominid specimens and stone tools was collected.

By studying the anatomy of the fossils, paleontologists began to distinguish the various species, but their guesses at the ages of the fossils remained not only relative, but speculative. Consider a comment made by Joseph Prestwich, an English geologist who had crossed the Channel to look at Boucher de Perthes' stone axes buried in the same strata as extinct animals. Prestwich immediately realized that this was incontrovertible proof of humankind's antiquity, and when he promptly published his conclusions he also vindicated John Frere's similar finding in 1800. Yet he also had this to say:

"The author does not, however, consider that the facts, as they at present stand, of necessity carry back Man in past time more than they bring forward the great extinct Mammals toward our own time, the evidence having reference only to relative and not to absolute time . . . "[7]

This problem still existed throughout the first half of the twentieth century.

If we could only guess that the Turkana boy might be several thousand or even several hundreds of thousands of years old, instead of *know* that he was a *Homo ergaster* preteen who lived 1.6 million years ago, what meaning would he have for us? If we thought the magnificent lions and horses of the Chauvet cave were the work of Celts and not of Ice Age tribes living over thirty thousand years old, the spell cast by this art would not be nearly so powerful. Speaking in the early nineteenth century of the newly revealed ages of Stone, Bronze, and Iron, a contemporary said "We know that [they are] older than Christendom, but whether by a couple of years or a couple of centuries, or even by more than a millennium, we can do no more than guess."[8] Even a few decades later these time periods were known to be far too short, but on the matter of guessing, the statement still held. Had it not been for the discovery of the radioactive clocks, we would still be guessing today.

Before we go any further, we need to put this "ancient" world of human prehistory into some kind of clearer perspective. Let's take the discoveries about our planet and ourselves which were made during the past two centuries, and put them into the order that the clocks have found for them during the century since their *own* discovery. If the 1,250 feet of the Empire State Building were made to stand for the 4.5 billion years since the earth's birth, then its first 102 stories, plus seventy-five feet of the shaft above, would represent all of the earth's time up until the Cambrian, the first period of the Paleozoic era of the geological column. The 155 feet of the remainder of the shaft would comprise all of the Paleozoic, Mesozoic, and

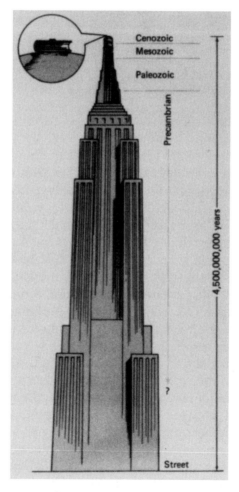

35. The Empire State Building used as a timescale for the earth's history.

nearly all of Cenozoic time. If a big book about ten inches thick were placed on top of the shaft, it would represent all of the time during which prehistoric humankind walked the earth. The time since the earliest civilizations arose in western Asia would be encompassed by a dime placed on top of that book, and a sheet of paper laid on top of the dime would tell the time elapsed since the birth of Jesus.[9]

All that we are dealing with now is the book—a period that covers only about five *million* years instead of the nearly nine hundred times greater 4.5 *billion* we have looked at already. This shift presents a problem for the radioactive clocks we have dealt with so far. They can still count most of the book, but when we get down to its last chapters, their very power makes them suddenly worthless. Their half-lives are too long; they simply can't cut time into tiny enough pieces because it takes too long to accumulate enough daughter product to count. And because the time counted by all radioactive clocks has a margin of error which is a percentage of the half-life, as smaller and smaller periods of time need to be distinguished, the margins of error of these powerful clocks become far greater than the periods of time which need to be counted!

With its 4.5 billion year half-life, the uranium-lead clock is much too powerful to distinguish bits of time within the range of hundreds of thousands of years, but the potassium–argon clock, with its half-life of "only" 1.3 billion years, works beautifully for dating most of the period of human prehistory comprised by the book balanced on the pinnacle of the Empire State Building. This clock burst upon the world in the 1950s, and by the next decade it had initiated a revolution in the study of our forebears by dating the first human fossil.

During the remaining decades of the century, this clock has opened up an element of communication for us with the silent skulls whose empty eye-sockets stare out at us so heart-stoppingly. The barrier between past and present remains, and we can never reach out to touch them or to speak to them, but by telling us when our forebears lived, the potassium–argon clock has done as much to bring them closer as has our growing knowledge about their stone tools or ways of life or powers of vocalization.

To give a sense of just how much a matter of guesswork dating was as late as 1960, consider the case of the *Australopithecus robustus* skull found in 1959 by Mary Leakey in the Olduvai Gorge of Tanzania's Serengeti Plain. The next year, Louis Leakey described it as more than 600,000 years old. Such dates, based on guesses about the time required for sediment to accumulate or evolution to proceed, had been given throughout the first half of the century, but they were known to have very little scientific value and were mostly produced for popular publications. [10]

The Olduvai Gorge, which holds the world's greatest treasure trove of human fossils, is as ideal for relative dating as any place could be. The Gorge cuts through a veritable layer cake of deposits, so that what lived before and after can easily be discerned, but such relative ages could bring us no closer to the absolute ages of our ancestors than the geological column, by itself, brought us to the true ages of earlier forms of life.

If scientists didn't pay much attention to Louis Leakey's date of 600,000 years, it was another matter altogether when just a few months later he corrected that date to a figure three times more ancient: 1.75 million years! Sediments associated with the fossil had been given this date by the potassium–argon clock developed by geologists and physicists at the University of California, Berkeley. Just a century after the publication of Darwin's *Origin of Species*, the first *absolute* date for a hominid fossil had been found.

By dating very young fine-grained volcanic rocks that can't be dated by any other method, the potassium–argon clock has been able to tell us about the young areas of our earth. Thanks to it, we now know that the Hawaiian Islands grew from a chain of volcanoes during a period from about five million to one million years ago.[11] As luck would have it, East Africa was also a volcanically active region during the period from one to four million years ago when the fossils were deposited. Layers of volcanic ash are sandwiched between those containing bones and artifacts, so that once the ash is dated by the potassium–argon clock, the ages of the bones and implements between the dated layers can be calculated. Without this clock, we couldn't know that Lucy is 3.18 million years old, or that the Turkana boy lived about 1.6 million years ago. In some cases, the signs of our ancestors are embedded directly into the volcanic ash. The famous 3.6 million-year-old *Australopithecus afarensis* footprints discovered by Mary Leakey at Laetoli, Tanzania in 1978 march across freshly deposited ash and give us irrefutable proof that these hominids walked upright.

The great story of human evolution that we have today resembles that of the geological column. Once again, our knowledge stems from utterly different disciplines: the new geology got a long timeframe accepted, archaeology's Three Ages of prehistory authenticated our ancient origins, and Darwin's concept of evolution motivated paleontologists to search for evidence even farther back into the depths of time. The picture they have created together could never have been painted by one of them alone. Had these discoveries not been made, and had the fossil hunters not been successful in finding the bones and identifying the species which preceded us, there would be no story. It is their work that lets us know that the Java man belonged to the *Homo erectus* species and the Turkana boy to the closely related *Homo ergaster*, while the much older Lucy was a member of one of our earliest ancestral species, *Australopithecus afarensis*. But without the temporal structure the clocks gave to the story, we could only have vaguely grasped the relationships among these species in time, and we would still be speculating about the ages of our earliest progenitors.

And there is something else. Without the clocks, our desire to see all evolutionary roads as leading to *us* might have prevented us from realizing that the story is not one of a single linear transformation, but rather of a tree with many branches, some of them coexisting over long periods of time. Thanks in large part to the clocks, we now know that the situation today of there being just one hominid species—*Homo sapiens*—throughout the world is the anomaly and not the norm. Between 2.5 and 1.5 million years ago, there were times when as many as *six* hominid species may have coexisted, and it now appears that less than fifty thousand years ago the earth was inhabited by *three* hominid species. We shared Europe with the Neanderthals and Southeast Asia with *Homo erectus*.[12] Without the neutral, quantitative decree of the radioactive clocks, we might never have accepted a reality so at odds with our own self-love.

There is also a different sort of clock which dealt yet another blow to our belief in our own singularity, and which cannot go unmentioned in the story of human evolution. Like the clocks based on the constancy of radioactive decay, the molecular clocks depend on the constancy of the rate of mutation in our DNA, the genetic code which determines the make-up of all the proteins in our bodies. Assuming a constant mutation rate, the degree of difference in the DNA of two species is directly related to the length of time since they diverged from a common parent stock. The first molecular clock was "invented" in 1967 by Allan Wilson and Vincent Sarich, also of the University of California, Berkeley, who compared certain blood proteins from living African apes and humans, ran the constant mutation rate backward in time, and found that apes and humans could not have diverged earlier than about five million years ago.

This result deviated wildly from the then current belief that the first hominid species had diverged from the apes at least as early as fifteen million years ago, and possibly as long ago as thirty million years. This long timeframe was based largely on the tenuous evidence of a partial upper jawbone with teeth found in Miocene strata. From this jawbone the existence of a hominoid species christened *Ramapithecus* had been conjured, complete with the attributes of bipedal locomotion, the ability to hunt rather than merely scavenge, and the habit of living in communal groups. Among the several causes for this very unscientific, non-empirical leap was surely the emotional motive of putting distance between the "special" human species and the rest of nature. If all of these human traits arose so long ago and so abruptly, then the comforting idea could be preserved that humans were a fundamentally different sort of creature from all others. So powerful was the hold of this belief that the radically contradictory evidence of the molecular clock was generally dismissed.[13]

Debate raged for over a decade, and it was not until the 1980s that the conflict was finally resolved in favor of the timeframe calculated by the molecular clock. Not only were more *Ramapithecus* fossils found which clearly showed that the species was an ape and not a hominid, but a large cache of fossils was discovered by Donald Johanson in the Afar region of Ethiopia which confirmed the existence of the ape-like hominid *Australopithecus afarensis* between three and four million years ago. Today, all signs point to this species as the stem from which the other hominid species branched.

The molecular clock not only gave us a far more recent, and thus more intimate, relationship with our primate cousins, but it compounded that offense by putting us on a kind of equal genetic footing with them by revealing that the proteins of chimpanzees and gorillas differed as much from each other as either did from human proteins. Although the time of the three-way split has now been pushed back from five to about seven million years, the molecular clock's results continue to hold the day. In fact, some researchers now believe

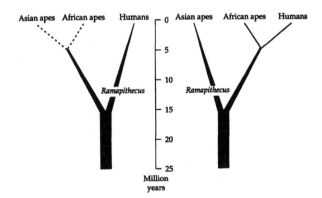

36. Before the discovery of the molecular clock in 1967, anthropologists interpreted the fossil evidence as indicating that humans had diverged from apes at least fifteen million years ago (left). The molecular clock revealed that the evolutionary divergence was far more recent: it now appears that hominids separated from the apes around seven million years ago (right).

the split was not three-way, and that humans and chimps actually have a closer genetic relationship to each other than either has with the gorillas.[14] Once again, it was the sheer, blind, neutral counting of a clock that brought us through to reality. Whatever form it takes, a clock can cut through time and find its truth in a way that nothing else can, and make of our desires and preconceptions so much vain puffery.

We have spoken all along about the importance of cross-checking the dates recorded by different radioactive clocks. Here, the molecular clock served as a kind of corroborative cross-check for the fossil dates found by the potassium–argon clock. The earliest hominid fossils yet found date to about four million years ago, which fits well with their evolutionary divergence as calculated by the molecular clock. Had we really diverged fifteen million years ago, it would be odd indeed that not a single hominid fossil has as yet been found nearer to that time.

The potassium–argon clock has dated volcanic rocks as young as 200,000 years, and some even as young as 100,000 years, but most of its reliable dates stop at around 500,000 years. Because of its long half-life, only occasionally can enough argon-40 be retrieved from rocks less than 500,000 years old so that an accurate measurement can be made. Thus, there is a period of nearly half a million years in our recent past, which is more difficult to date than any other.[15] Radioactive clocks which use shorter half-lifed intermediate members of the uranium-lead decay series are helping to create a temporal framework for this period, as are a host of other methods which can interlock with one another to build up a picture. They can also be used to cross-check the readings of the potassium–argon clock as well as those of each other.

The technique of fission-track dating counts the damaged zones, or tracks, etched into a crystal when uranium-238 undergoes spontaneous fission, i.e., when it splits into two more stable elements, releasing energy in the process. Assuming equal amounts of parent material, the more fission tracks formed, the older the crystal. Two other techniques, thermoluminescence (TL) and electron spin resonance (ESR), are based on the fact that electrons accumulate over time as radiation from radioactive elements in rocks continually punctures the crystal lattice. Thermoluminescence can determine the age of certain radioactive minerals by reheating them and then counting the bits of light emitted as electrons escape from the crystal lattice in which they have been entrapped. Electron spin resonance is also able to date minerals by counting the number of electrons trapped in crystals after radioactive decay—it counts them not by the light they emit when heated, but by exploiting their ability to absorb microwave radiation. Because electrons accumulate with time, the number of electrons counted indicates the age. TL is able to date many stone tools and ceramics, while ESR works well on tooth enamel, shells, and coral.

In conjunction with these and other dating methods, another brilliant technique based on the atom has allowed us to chart the changing climates of this period of recurring Ice Ages. Like the radioactive clocks, the nonradioactive oxygen-18/oxygen-16 system uses the mass spectrometer's ability to distinguish the infinitesimal differences in mass of isotopes. Ninety-nine percent of the earth's oxygen is oxygen-16, and most of the remainder is oxygen-18. When water evaporates, the lighter molecules containing oxygen-16 evaporate more readily than those containing oxygen-18. Normally, the water which evaporates from the surface of the sea just condenses and returns to the ocean as rain or river water, so that the ratio of oxygen-16 to oxygen-18 in the sea remains the same. But during an Ice Age, the evaporated water becomes trapped in ice sheets and doesn't return, so that the proportion of oxygen-18 in the sea increases. Skeletons of sea animals found in cores drilled deep into ancient sediments beneath the sea have the same proportion of oxygen-16 to oxygen-18 as existed in the sea which surrounded them in life. Thus, the ratio of oxygen-16 to oxygen-18 in these fossils tells us whether or not there was an Ice Age in progress at the time they were formed.[16] This information has been correlated with dates given by the radioactive clocks to reveal the climate changes throughout the last few hundred thousand years.

Our picture of the past half million years is slowly coming into focus, but the closer we get to our own moment in time, the more we desire a detailed picture of the events which brought us here. What of the time closest to us—say, the past 50,000 years or so? This period—which requires deciphering time at the level of thousands of years, cannot be well dated by any of the methods described so far. It is as nothing in the grand scheme of our planet—just a tiny bit of the book plus the dime and the piece of paper in the Empire State Building analogy. It is of no consequence at all—except to *us*. Consider only a few of the events it witnessed: the last Ice Age, the lives and extinctions of our last rival hominid species, the trek of Asians across the Bering Strait to the western hemisphere, the extinctions of such mammals as the mastodons, woolly mammoths, and saber-toothed cats . . . Then, closer to home, the advent of farming and cities and the rise of metallurgy . . . Troy, Stonehenge, the pyramids and the Parthenon.

Many of the events encompassed by the dime and the sheet of paper are datable by written records, but certainly not all. The great civilization of the Inca had no writing, and every year villages are discovered, and pottery and artifacts unearthed from we know not when. Where do these belong in the calendrical timeline we have put together from written records? As we will see in the next chapter, only the discovery of another great clock could help us find the answer.

Inconsequential in terms of the whole of time this period may be, and yet it was only the people on the very top surface of the sheet of paper who—by discovering radioactivity and its clocks—have given us the perspective without which we couldn't be aware of our own inconsequence.

Chapter XXIII

The Clock That Came from Outer Space

Children picking up our bones
Will never know that these were once
As quick as foxes on the hill; . . .
And least will guess that with our bones
We left much more, left what still is
The look of things, left what we felt
At what we saw. . . .

Wallace Stevens
A Postcard from the Volcano

Space is simply there—clear and unproblematical. The wide plains thrill us, the mountain heights awe us. Only the incomprehensible distances of the universe send a shudder down our spines. Time is just as objective a phenomenon as space, and yet we humans cannot regard it with the same plain joy that we extend to space. It is problematical and disturbing for us in a way that space is not. Time brings us pressure, and boredom, and above all, it brings us our own deaths.

The great divide between life and death which we experience in our personal lives extends to history, and brings us the sorrow we feel at past cultures gone forever. We are of two minds about the past. The thoughtless, life-loving part of us can never quite believe that our forebears were as *real* as we are. But in our hearts we know the truth: they were once every bit as alive as we are now, and that awareness pricks the fragile skin holding us apart from our own mortality, and makes us yearn to know more about those who came before.

The past is a great sea. We want to know about it all, but we hunger most for knowledge of those last frothy bits of foam that lap closest to the shore we now inhabit. We found the debris left by the waves—the shells and bones and pots and bits of wood—but unless we could learn *when* they were washed up, the story these remnants could tell would remain jumbled and inchoate.

Couldn't some kind of clock be found which could help give us a sense of the history and lives of our ancestors during the last fifty thousand years or so? Relative to those we have looked at so far, it would have to be a clock that didn't count hours, or even minutes, but only seconds. It would have to be a clock that could cut time up into much smaller pieces than can those we have already considered. In short, it would have to be a clock with a very short half-life.

Throughout the first half of the twentieth century, there was not even a glimmer of a hope that such an odd second hand clock could be found. Yet by the 1940s it was known that carbon-14, a radioactive isotope of stable carbon-12, could be produced in the laboratory by bombarding nitrogen atoms with neutrons. The neutron is absorbed by the nitrogen atom's nucleus, which causes it to emit a proton. Since the number of protons determines the element, the emission of that single proton turns the nitrogen atom into a carbon atom. (Carbon atoms have six protons rather than nitrogen's seven.)

Like all other radioactive substances, carbon-14 decays at a constant rate, and its half-life has been calculated to be a mere 5,730 years. Just as with the other clocks, in a population of carbon-14 atoms, half of them will decay into their stable daughter substance in 5,730 years, and then half of the remaining in another 5,730 years, and so on. In this case, the stable daughter is nitrogen-14, from which radioactive carbon-14 was originally made. As it reverts to nitrogen-14, carbon-14 emits an electron. It was by counting the emissions in a given period of time that the half-life was determined.

So a new clock was discovered in the laboratory, but what use was it? There are many more radioactive clocks than we have mentioned, but they don't work well for radiometric dating: it may be that they are simply too rare, or that their half-lives are so long that sufficient countable daughter substance can't build up, or alternatively, their half-lives may be so short that too soon no parent substance is left to count. The radiocarbon clock itself can't give us dates back much farther than 50,000 years ago, because at that point there is so little carbon-14 left that it can no longer be counted. Limited it may be, but it is limited to a period of time that is very important to us.

This clock inspires our awe in a different way from the long half-lived clocks which carried us over the eons to the dawn of the earth. Carbon-14 can't do anything like that, and yet this little second hand clock binds us part and parcel to our planet, and even to the universe, like nothing else. Unlike the other clocks, it enters into the biosphere, and thus into *us*. Carbon is one of the most common elements in living things—to say that the radiocarbon clock can date anything that has carbon in it is to say that it can date nearly everything that was once alive: wood, charcoal, grains, seeds, grasses, cloth, paper, hides of animals, charred bones, peat, ivory, shells, carbon-containing pottery, and more.

Almost as awe-inspiring as the clock itself was the brilliant insight of Willard F. Libby (1908–1980) which gave it its birth, and for which he won the Nobel Prize. Libby looked at the transformation of nitrogen into radioactive carbon in the laboratory and realized that the same thing must be going on in the earth's upper atmosphere all the time. As we will see, these last three words "all the time" held the key to transforming this little clock into a powerhouse.

At altitudes of five to ten miles, the earth's upper atmosphere is continually being bombarded by high velocity cosmic rays. These rays are really individual atoms produced by the solar flares which rise millions of miles above the sun's surface. They have traveled ninety-three million miles to earth, where they collide with atoms in the outer atmosphere with all the force that man can produce with a particle accelerator. Free neutrons result from these collisions, which in their turn collide with and are captured by nitrogen atoms, causing them to each emit a proton and thus transform themselves into unstable, radioactive carbon-14 atoms, just as happens in the laboratory.

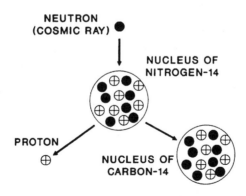

37. Cosmic rays change nitrogen to carbon-14 in the upper atmosphere. Radioactive carbon-14 slowly decays back to stable nitrogen-14.

Like their stable carbon-12 sisters, most carbon-14 atoms don't remain free, but combine with oxygen atoms to make molecules of carbon dioxide (CO_2). As the winds churn up the atmosphere, these radioactive CO_2 molecules make their way into the lower atmosphere and are eventually taken in by plants in the process of photosynthesis. Human beings eat the plants and also eat animals who have eaten the plants. Carbon is one of the basic building blocks of life and occurs in all the amino acids which make up the proteins of our bodies. *All* of it ultimately comes from the atmosphere. That not only binds us to the whole of nature, but it also makes us, along with everything else, just a little bit radioactive. How little? The ratio of radioactive carbon-14 to stable carbon-12 remains the same whether it is in the atmosphere or in a plant, animal or human, and that ratio is about one carbon-14 atom for every *trillion* carbon-12 atoms.

After fifty thousand years, only one fifth of one percent (two parts out of every thousand) of the original carbon-14 remains in any sample of natural carbon.[1] In light of the fact that the original carbon-14, at its *maximum*, only made up one part in a trillion of the carbon in living things, this is a mind-bogglingly small amount. Once again, it is the gargantuan quantities of atoms (and the extraordinary technology of our measuring instruments) that allow such minute percentages to still be countable. But beyond about fifty thousand years, there simply isn't enough left to count.

There's no need to ask why no one had ever detected carbon-14 in either the atmosphere or in living things. Libby set out to find this infinitesimal amount, and he succeeded. Then, in 1948, he showed how the carbon-14 in the atmosphere could be used as a radioactive clock. This clock is exactly like the others in that it is based on the constancy of its isotope's decay rate, but it works a little differently. The age of carbon-bearing material isn't determined by the parent–daughter ratio as in the other clocks, but by the ratio of carbon-14 to all of the other carbon in the sample.

Here's how it works. Carbon-14 in the atmosphere is essentially in equilibrium—the rate of production is equal to the rate of decay. The ratio of one carbon-14 atom to one trillion carbon-12 atoms remains the same in the atmosphere, in the plants that take in CO_2, and in the animals that eat the plants. As carbon-14 atoms decay, an equal amount of carbon-14 arrives from the outer atmosphere to replenish the stock flowing into plants, animals, and *us*. It is this constant replenishment that allows this clock to be used so successfully for radiometric dating.

When a plant or animal dies, that replenishment is cut off. No new carbon will ever be absorbed by it. This is the point at which the radiocarbon clock starts ticking. Although dead organic matter has no more interactions with its environment, the carbon-14 already inside it keeps on decaying at its constant rate. Now, for the first time, the ratio of radioactive carbon-14 to stable carbon-12 begins to decline. Let's take as an example a tree in Wisconsin which was uprooted and killed when an ice sheet bulldozed over it sometime in the past. When the ratio of carbon-14 to carbon-12 in a sample of this tree's dead wood was determined in the laboratory, it was found to be far lower than the ratio of one carbon-14 atom to one trillion carbon-12 atoms found in the atmosphere and in all living things. Knowing that carbon-14's half-life is 5,730 years, scientists then compared the two ratios and determined that the tree was 11,400 years old. We are thus able to infer that an ice sheet overran a Wisconsin forest as recently as 11,400 years ago.[2]

By now we have figured out that no clock can escape some problems, and the radiocarbon clock is no exception. The accuracy of this clock is based on the assumption that the level of carbon-14 in the atmosphere has remained constant at least throughout the period that the clock can measure. Unfortunately, we have now found that this is not quite true. Fluctuations in the sun's solar flares may change the intensity of cosmic ray bombardments of the upper atmosphere, which in turn causes variations in carbon-14 production. Thus, calculations based on today's level of carbon-14 in the atmosphere will not be quite accurate.

Fortunately, for the relatively recent past there is a remedy: A precise chronology of actual years has been laboriously built up using tree rings. It is based on the rings of many trees, including the very ancient bristle-cone pines and sequoias. The rings have been overlapped and correlated, and the unbroken chronology now extends back to about 7000 B.C. When radiocarbon dates for rings of a certain age are plotted against the real ages, the deviation due to carbon-14 fluctuations is revealed. This can then be used to correct the radiocarbon dates for samples of other materials around the same age. Dates earlier than 7000 B.C. must of course remain uncalibrated.[3]

The radiocarbon clocks can also give inaccurate readings because of a problem we have met with many times before: the system is not closed to the outside. An old porous bone or piece of wood that is buried in the earth can pick up new carbon from plant rootlets that grow into it. If this carbon were included in the sample, it would make it appear to be more recent than it actually is. There are now good techniques for removing this contamination and it is rarely a problem in material less than twenty-five thousand years old.

Clearly, the radiocarbon clock doesn't have quite the accuracy that archaeologists would like it to have, but it can go where nothing else can, and it has opened up the past fifty thousand years to us. When a village is uncovered far down in South America, its charcoal, seeds, and bits of dead animal matter can be dated to tell us about the advance of people into the western hemisphere. New dates are being determined every year, and the story of our migrations not only here but throughout the world is in a state of seemingly constant revision. With each new date, we learn not only *when* but *what*. When seeds of domesticated plants are dated, we know that agriculture had been mastered in that place by that time, and the same goes for the dating of dead organic matter from domesticated animals. The date for a bit of cloth lets us know that the technology of weaving existed by then. More often than not, a new piece of knowledge wrested from the past by the radiocarbon clock doesn't corroborate old knowledge, but instead wrenches us away from our presuppositions and forces us to view things from a new perspective. In this way, slowly but surely, date by date, we are substituting fact for belief and constructing at least the lineament of a map of our true past.

A case in point involves the prehistoric cave paintings in France and Spain. Charcoal, which is simply burnt wood, has been one of the most fruitful materials for radiocarbon dating. Not only charcoal found on the floors of caves but also (since the advent of the accelerator mass spectrometer) tiny bits of charcoal taken from the paintings themselves have revealed the Altamira art to be fourteen to sixteen thousand years old, that of Lascaux about seventeen thousand years old, and the gorgeously naturalistic horses and lions of the recently discovered Chauvet cave to be around 31,000 years old.

Just as the first radioactive clocks demolished the nineteenth century theories about the age of the earth, so this date for the Chauvet art has turned on its head the conviction, unchallenged during most of the twentieth century, that the cave art had progressed from rather stiff, crude beginnings in the earliest work toward ever greater sophistication in drawing technique, perspective, and coloration as the millennia wore on. The Chauvet Cave is not only the earliest yet dated, but its magnificent, mobile animals are among the most accurately drawn and exquisitely colored of all the creatures found in any of the caves. Once again, the blind counting of the clocks has brought us to a truth more fascinating than any theory we might weave from our own prejudices. The myth of a slow linear artistic evolution in the cave art allowed us to distance ourselves from our supposedly rather crude and dimwitted ancestors. Now the radiocarbon dating of Chauvet has forced us to realize that these forebears may have lacked our knowledge and technology, but they were still *us* through and through, and they really did make these great paintings over thirty thousand years ago.[4]

Radiocarbon clocks have demolished other cherished preconceptions as well. They have shown that stone temples were built in Malta before the pyramids, that crops were cultivated beside the Rhine sooner than along the Nile, and that a buried city in Turkey predates any in the Mesopotamian Fertile Crescent. "Radiocarbon rescued archaeology from the circular reasoning that assigned relative dates according to how advanced a culture seemed to be."[5]

* * *

Thousands upon thousands of objects have already been dated by the radiocarbon clock, and thousands more will be in the future. Perhaps someday another T-shaped piece of wood with some notches on it will be dug up, perhaps in a country other than Egypt. And if this newly-discovered gnomon is found somewhere else and is radiocarbon-dated to earlier than about 1500 B.C., we will have to revise our ideas about the origin of our earliest efforts to tell the time of day. Whether or not such a discovery is made, with thoughts on the first scientific instrument for measuring time, our present story has come full circle.

The very first effort to measure time was of course the counting from new moon to new moon, which may have begun around the time the first cave art was being painted.[6] Then came the Egyptians' discovery of the solar year, which they found by watching the circling of the stars overhead and counting the days between the appearance of the dog star Sirius as the last visible star above the horizon just before dawn. Finding the precise length of the solar year was difficult, but like the observation of the phases of the moon, it required no instruments. With the gnomon, humankind entered a new phase in its effort to gain control over time by measuring identical segments of it. We no longer simply counted the signs given us by nature; we devised our own. We invented the time of day as much as we discovered it.

During the following millennia, many cultures developed unique conventions by which they strung years together to make continuous chronologies that could help them hold onto their pasts, but the most ingenious temporal enterprise of that long period was the development of ways to mechanically count the evanescent time of day. In all that long time, this is as far as we got in mapping the realm of time. Had it not been for the discovery of the atom and its radioactivity, this is where we would have remained.

What if a catastrophe occurred tomorrow like the collision of the giant meteor which decimated the dinosaurs sixty-five million years ago? And what if a handful of *Homo sapiens* survived, and slowly, over millennia, built themselves back up to the point where they began to excavate in order to discover their own prehistory? In reality, a layer of sediment comprising only a century could probably never be found, but let us imagine that they were able to locate a twentieth century stratum. What would strike them most about its human fossils would not be our beads and buckles, but something else: in our layer of sediment there would be a metal instrument with numbers on it near the lower end of almost every fossil arm bone. They might label the fossils found in this narrow stratum the people of the wrist watch.

If they later became able to decipher our script, and if they were fortunate enough to stumble upon some scientific writings about radiometric dating, it is just conceivable that they could figure out we were not only the people of the wrist watch, but also the people who had found the key to unlock the depths of time. If they realized that the wrist watch was an instrument to tell its wearer the precise present moment, then they would know we were at once the people of the present moment and the people who could look the farthest out from the moment in which they were entrapped.

It is unlikely that our descendants would find the remains of the atomic clocks which lie like a beating heart at the center of all our communications, energy, and transportation systems, but if they did, they would discover that the atom was the key to both the present and the past in the twentieth century. Had the decimating cataclysm occurred just a hundred years earlier, at the end of the nineteenth century instead of the twentieth, our descendants would have uncovered neither wrist watches nor atomic clocks nor radioactive clocks.

Even the most scrupulous culling of the fossil record in the layers beneath the twentieth century stratum would never yield the long, complex story of how we finally arrived at telling both present and past time by the constant time segments generated by subatomic particles. Like that tale, there is another which would be forever lost to those who came after. As they pondered the question themselves, these future archaeologists might hope that someday, somewhere, they would unearth from our layer of sediment the answer to what the thing itself *is*. In that hope, they would be disappointed. For all our ability to so exquisitely measure it, *what time is* remains as much a mystery for us as it was for St. Augustine in the fourth century A.D. Right along with him we still ask:

"For what is time? . . . Who can even in thought comprehend it, so as to utter a word about it? . . . If no one asks me, I know: If I wish to explain it to one that asketh, I know not. . . ."

References

Part One: The Time of Day

Introduction

1. *The Confessions of St. Augustine* (New York: Random House [The Modern Library] 1949), Book XI, p. 253.

Chapter 1. The Planetary Basis of Our Day

1. Neil F. Comins, *What If the Moon Didn't Exist?* (New York: HarperCollins, 1993), pp. 7–13.
2. Ibid., pp. 5–6.
3. William K. Hartmann and Ron Miller, *The History of Earth: An Illustrated Chronicle of an Evolving Planet* (New York: Workman, 1991), p. 54.
4. Comins, *What If the Moon Didn't Exist?*, pp. 61–62.
5. Sam Flamsteed, "Impossible planets." *Discover*, 18:9, 1997, p. 78.
6. Comins, *What If the Moon Didn't Exist?*, p. 65.
7. G. Brent Dalrymple, *The Age of the Earth* (Stanford, CA: Stanford University Press, 1991), pp. 51–52.
8. Charles A. Schweighauser, *Astronomy from A to Z: A Dictionary of Celestial Objects and Ideas* (Springfield, IL: Illinois Issues, 1991), p. 95.
9. Isaac Asimov, *Asimov's New Guide to Science* (New York: Basic Books, 1972), p. 98.
10. Hartmann and Miller, *The History of Earth*, p. 71.
11. Comins, *What If the Moon Didn't Exist?*, pp. 53–54.
12. Malcolm W. Browne, "New gauge for shorter day of past." *The New York Times*, July 9, 1996, p. C6.
13. Asimov, *Asimov's New Guide to Science*, p. 176.
14. Gail S. Cleere, "Making Time." *Natural History* 6/94, p. 86.
15. Asimov, *Asimov's New Guide to Science*, p. 176.

Chapter II. The Sundial and Its Temporary Hours

1. G. J. Whitrow, *Time in History: Views of Time from Prehistory to the Present Day* (Oxford: Oxford University Press, 1988), p. 15.
2. Homer, *The Iliad*, as quoted in Whitrow, p. 15.
3. Homer, *The Odyssey*, trans. E.V. Rieu (Baltimore, MD: Penguin Books, 1946).
4. Whitrow, *Time in History*, p. 28.
5. Jerome Carcopino, *Daily Life in Ancient Rome* (New Haven, CT: Yale University Press, 1940), p. 147.
6. Ibid., pp. 145–146.
7. Ibid., p. 150.
8. Ibid., pp. 150–151.
9. Ibid., p. 145.

Chapter III. The Measurement of the Night Hours

1. G.J. Whitrow, *Time in History* (Oxford: Oxford University Press, 1988), p. 91.
2. Carcopino, *Daily Life in Ancient Rome* (New Haven, CT: Yale University Press, 1940), p. 147–148.
3. Silvio A. Bedini, "Clocks and the reckoning of time," in *Dictionary of the Middle Ages*, ed. Joseph R. Strayer (New York: Charles Scribner's Sons, 1982), p. 458.
4. Daniel J. Boorstin, *The Discoverers* (New York: Vintage Books, 1985), p. 29.
5. Sigvard Strandh, *The History of the Machine* (New York: Dorset Press, 1989), p. 170.
6. Bedini, "Clocks and the reckoning of time," p. 458.
7. Marc Bloch, *Feudal Society* (Chicago: The University of Chicago Press, 1961), p. 73.

Chapter IV. The Canonical Hours

1. David S. Landes, *Revolution in Time: Clocks and the Making of the Modern World* (Cambridge, MA: Harvard

University Press, 1983), p. 59. For discussion, see pp. 58–62.

2. Ibid., p. 60.

3. *St. Benedict's Rule for Monasteries* (Collegeville, MN: The Liturgical Press, 1948), Chapters 8–19, pp. 29–41.

4. Silvio A. Bedini, "Clocks and the reckoning of time," in *Dictionary of the Middle Ages*, ed. Joseph R. Strayer (New York: Charles Scribner's Sons, 1982), p. 457.

5. R. Poole, "A monastic star timetable of the 11th century" in the Journal of Theological Studies, 1914, XVI, pp. 98–104, as quoted in R.W. Southern, *The Making of the Middle Ages* (New Haven: Yale University Press, 1953), p. 187.

6. Southern, *The Making of the Middle Ages*, p. 187.

7. Jacques Le Goff, "Merchant's time and church's time in the Middle Ages," in *Time, Work and Culture in the Middle Ages*, by Jacques Le Goff (Chicago, IL: The University of Chicago Press, 1980), p. 38.

Chapter V. The Selling of Time

1. *Tabula exemplorum* (ed. J.T. Welter [1926], p. 139), as quoted in Jacques Le Goff, *Time, Work and Culture in the Middle Ages* (Chicago, IL: The University of Chicago Press, 1980), note 2, p. 290.

2. Norman F. Cantor, *Civilization of the Middle Ages* (New York: HarperCollins, 1963), p. 480.

3. Ibid., p. 128.

4. Le Goff, "Licit and illicit trades in the medieval west," in *Time, Work and Culture in the Middle Ages*, p. 61.

5. R.R. Palmer, *A History of the Modern World* (New York: Alfred A. Knopf, 1959), p. 29.

6. Dante, *The Inferno*, trans. John Ciardi (New York: The New American Library, 1954), Canto XVII, Circle 7, Round 3, p. 151.

7. Cantor, *Civilization of the Middle Ages*, p. 365.

8. Ibid., p. 228.

9. Palmer, *A History of the Modern World*, p. 98.
10. Le Goff, "Trades and professions as represented in confessors' manuals" in *Time, Work and Culture in the Middle Ages*, pp. 107–121.
11. Le Goff, "Labor time in the 'crisis' of the fourteenth century," in *Time, Work and Culture in the Middle Ages*, p. 48.
12. Ibid., p. 38.

Chapter VI. The Mechanical Clock: The Product

1. Lewis Mumford, *Technics and Civilization* (London: Routledge & Kegan Paul, 1934), p. 14.
2. Ibid., p. 134.
3. David S. Landes, *Revolution in Time* (Cambridge, MA: Harvard University Press, 1983), p. 6.
4. Mumford, *Technics and Civilization*, p. 15.
5. Isaac Asimov, *The Clock We Live On* (Eau Claire, WI: E.M. Hale and Co., 1968), p. 29.
6. Mumford, *Technics and Civilization*, p. 15.
7. Daniel J. Boorstin, *The Discoverers* (New York: Vintage Books, 1985), p. 13.
8. G.J. Whitrow, *Time in History* (Oxford: Oxford University Press, 1988), p. 108.
9. Mumford, *Technics and Civilization*, p. 17.
10. Ibid., p. 15.
11. Ibid., p. 16.
12. John Boslough, *Masters of Time: Cosmology at the End of Innocence* (Reading, MA: Addison–Wesley Publishing Co., 1992), p. 175.

Chapter VII. The Mechanical Clock: The Machine

1. Lynn T. White, Jr., *Medieval Technology and Social Change* (Oxford: Clarendon Press, 1962), p. 89.
2. Ibid., p. 119.

3. David S. Landes, *Revolution in Time* (Cambridge, MA: Harvard University Press, 1983), p. 8.
4. As quoted in Silvio A. Bedini, "Clocks and reckoning of time," in *Dictionary of the Middle Ages*, ed. Joseph R. Strayer (New York: Charles Scribner's Sons, 1982), p. 459.
5. H. Alan Lloyd, "Timekeepers—an historical sketch," in *The Voices of Time*, ed. J.T. Fraser (New York: George Braziller, 1966), p. 392.
6. Landes, *Revolution in Time*, p. 56.
7. G.J. Whitrow, *Time in History* (Oxford: Oxford University Press, 1988), p. 102.
8. Landes, *Revolution in Time*, pp. 67–68.
9. Ibid., pp. 56–57.
10. A replica of Giovanni de' Dondi's instrument exists in the Smithsonian Institute, Washington, DC, and replicas of both the Dondi and Wallingford machines can be found at the Time Museum, Rockford, IL.
11. Jean Gimpel, *The Medieval Machine: The Industrial Revolution of the Middle Ages* (New York: Penguin Books, 1976), p. 160.
12. Ibid., pp. 153–154.
13. Ibid., p. 159.
14. Landes, *Revolution in Time*, pp. 54–58.

Chapter VIII. The Early Machines

1. Jean Gimpel, *The Medieval Machine* (New York: Penguin Books, 1976), p. 168.
2. Daniel J. Boorstin, *The Discoverers* (New York: Vintage Books, 1985), p. 39.
3. Gimpel, *The Medieval Machine*, p. 169.
4. W. Rothwell, "The hours of the day in medieval French," *French Studies* 13 (July 1959), p. 242. Cited in David S. Landes, *Revolution in Time* (Cambridge, MA: Harvard University Press, 1983), p. 409.
5. Landes, *Revolution in Time*, p. 80.
6. Ibid., p. 81.

7. G.J. Whitrow, *Time in History* (Oxford: Oxford University Press, 1988), p. 110.
8. John Hunter, *Clocks: An Illustrated History of Timepieces* (New York: Crescent Books, 1991), pp. 51–52.
9. Gerhard Dohrn-van Rossum, *History of the Hour: Clocks and Modern Temporal Orders* (Chicago: University of Chicago Press, 1996), p.121.
10. Landes, *Revolution in Time*, p. 89.
11. Ibid., p. 205.

Chapter IX. The Pendulum

1. James Jesperson and Jane Fitz–Randolph, *From Sundials to Atomic Clocks: Understanding Time and Frequency* (New York: Dover, 1977), p. 27.
2. David S. Landes, *Revolution in Time* (Cambridge, MA: Harvard University Press, 1983), p. 119.
3. Isaac Asimov, *The Clock We Live On* (Eau Claire, WI: E.M. Hale and Co., 1968), p. 24.
4. John Hunter, *Clocks: An Illustrated History of Timepieces* (New York: Crescent Books, 1991), p. 28.
5. Landes, *Revolution in Time*, p. 124.
6. Ibid., pp. 122–123.

Chapter X. The Longitude: Where Where Becomes When

1. John Boslough, "The enigma of time," *National Geographic* 177:3, 1990, p. 127.
2. Daniel J. Boorstin, *The Discoverers* (New York: Vintage Books, 1985), p. 151.
3. Lloyd A. Brown, *The Story of Maps* (New York: Dover, 1949), pp. 50–53.
4. Boorstin, *The Discoverers*, p. 48.
5. G.J. Whitrow, *Time in History* (Oxford: Oxford University Press, 1988), p. 139.

6. Boorstin, *The Discoverers*, p. 235.
7. Whitrow, *Time in History*, p. 139.
8. For the story of how the Jovian moon method was used in mapping on land, see Chapter 8, "The family that mapped France" in John Noble Wilford's *The Mapmakers* (New York: Random House, 1981).
9. Ibid., p. 137.

Chapter XI. Of Time and the Railroad

1. G.J. Whitrow, *Time in History* (Oxford: Oxford University Press, 1988), p. 158.
2. Lawrence Wright, *Clockwork Man* (New York: Horizon Press, 1968), p. 145.
3. Michael O'Malley, *Keeping Watch: A History of American Time* (New York: Viking Penguin, 1990), p. 138.
4. Wright, *Clockwork Man*, p. 145.
5. Derek Howse, *Greenwich Time* (Oxford: Oxford University Press, 1980), p. 113.
6. Ibid., p. 120.
7. O'Malley, *Keeping Watch*, pp. 66–67.
8. Ibid., p. 74.
9. Ibid., p. 81.
10. Howse, *Greenwich Time*, pp. 227–228.
11. Excerpt from speech by W.F. Allen at the October, 1994, International Meridian Conference, as quoted in Howse, *Greenwich Time*, p. 145.
12. *Harper's Weekly*, Dec. 29, 1883, p. 843, as quoted in Howse, *Greenwich Time*, p. 126.
13. *Louisville Courier–Journal*, Nov. 22, 1883, p. 4, as quoted in O'Malley, *Keeping Watch*, p. 134.
14. *Pittsburgh Dispatch*, Dec. 28, 1886, p. 4, as quoted in O'Malley, *Keeping Watch*, p. 135.
15. *Cincinnati Commercial Gazette*, Nov. 24, 1883, p. 4, as quoted in O'Malley, *Keeping Watch*, p. 136.

Chapter XII. The Worldwide Web of Time

1. Derek Howse, *Greenwich Time* (Oxford: Oxford University Press, 1980), p. 141.
2. Ibid., p. 134.
3. Ibid., p. 135.
4. Ibid., p. 156.
5. Michael O'Malley, *Keeping Watch: A History of American Time* (New York: Viking Penguin, 1990), p. 109–110.

Chapter XIII. A Watch on Every Wrist

1. David S. Landes, *Revolution in Time* (Cambridge, MA: Harvard University Press, 1983), p. 287.
2. Ibid., p. 353.
3. John Boslough, "The enigma of time," *National Geographic* 177:3, 1990, p. 115.
4. Landes, *Revolution in Time*, pp. 227–229.
5. Ibid., pp. 300–302.
6. Ibid., p. 316.
7. Michael O'Malley, *Keeping Watch: A History of American Time* (New York: Viking Penguin, 1990), p. 172.
8. Ibid., pp. 183–184.
9. Neil Baldwin, *Edison: Inventing the Century* (New York: Hyperion, 1995), pp. 135–140.
10. O'Malley, *Keeping Watch*, p. 161.
11. Landes, *Revolution in Time*, p. 340.
12. Isabella de Lisle Selby, *Wrist Watches* (Philadelphia, PA: Courage Books, 1994), p. 26.

Chapter XIV. New Vibrations

1. Derek Howse, *Greenwich Time* (Oxford: Oxford University Press, 1980), pp. 176–177.
2. Samuel A. Goudsmit and Robert Claiborne, *Time* (New York: Time–Life Books, 1966), p. 102.

3. H. Arthur Klein, *The World of Measurements* (New York: Simon and Schuster, 1974), p. 170.
4. Ibid., p. 172.
5. James Jesperson and Jane Fitz–Randolph, *From Sundials to Atomic Clocks: Understanding Time and Frequency* (New York: Dover, 1982), p. 40.
6. Gary Taubes, "A clock more perfect than time," *Discover* 12/96, p. 72.
7. Howse, *Greenwich Time*, p. 182.
8. Jesperson and Fitz–Randolph, *From Sundials to Atomic Clocks*, p. 55.
9. Bill Klepczynski, as quoted in Cleere, "Making time," *Natural History* 6/94, p. 86.
10. Taubes, "A clock more perfect than time," *Discover* 12/96, p. 70.

Chapter XV. The Clock and Charles Darwin

1. Samuel Butler, *Erewhon and Erewhon Revisited* (London: J. M. Dent & Sons, Ltd., 1979), p. 44.

Part Two: The Time of the Earth

Chapter XVIII. The Circle and the (Short) Line

1. Mircea Eliade, *The Myth of the Eternal Return* (Princeton, NJ: Princeton University Press, 1954), p. 114.
2. Linda Schele and David Freidel, *A Forest of Kings: The Untold Story of the Ancient Maya* (New York: William Morrow, 1990), p. 82.
3. Timothy Ferris, *Coming of Age in the Milky Way* (New York: Doubleday Anchor Books, 1988), pp. 219–220.
4. Ibid., p. 220.
5. Stephen Jay Gould, "The fall of the house of Ussher." In *Eight Little Piggies: Reflections on Natural History*, by Stephen Jay Gould (New York: W.W. Norton & Co., 1993), pp. 188–189.

6. Stephen Toulmin and Jane Goodfield, *The Discovery of Time* (New York: Harper & Row, 1965), p. 56.
7. Robin Lane Fox, *Pagans and Christians* (New York: Alfred A. Knopf, 1989), p. 267.
8. Toulmin and Goodfield, *The Discovery of Time*, p. 72.
9. Marjorie Hope Nicolson, *The Breaking of the Circle: Studies in the Effect of the "New Science" upon Seventeenth Century Poetry* (New York: Columbia University Press, 1960), p. 109.
10. G. Brent Dalrymple, *The Age of the Earth* (Stanford, CA: Stanford University Press, 1991), p. 2.

Chapter XVIII. Discovering the "Dark Abyss"

1. Don L. Eicher, *Geologic Time* (Englewood Cliffs, NJ: Prentice–Hall, Inc., 1968), p. 19, and Donald R. Prothero, *Interpreting the Stratigraphic Record* (New York: W.H. Freeman & Co., 1990), p. 21.
2. Richard Morris, *Time's Arrows: Scientific Attitudes toward Time* (New York: Simon & Schuster, 1984), pp. 68–70.
3. Leonardo da Vinci, Leicester MS, folio 9 verso, as quoted in W.N. Edwards, *The Early History of Paleontology* (London: British Museum, 1976), p. 17.
4. Yvette Gayrard–Naly, *Fossils: Evidence of Vanished Worlds* (New York: Harry N. Abrams, 1994), p. 23.
5. Frank Dawson Adams, *The Birth and Development of the Geological Sciences* (New York: Dover Publications, Inc., 1938), p. 262.
6. Prothero, *Interpreting the Stratigraphic Record*, p. 6.
7. Claude C. Albritton, Jr., *The Abyss of Time* (San Francisco, CA: Freeman, Cooper and Co., 1980), p. 50.
8. Daniel J. Boorstin, *The Discoverers* (New York: Vintage Books, 1985), p. 456.
9. Francis Darwin, ed. *Charles Darwin's Autobiography* (New York: Schuman, 1950), p. 57, as quoted in Timothy

Ferris, *Coming of Age in the Milky Way* (New York: Doubleday Anchor Books, 1988), p. 235.

10. Toulmin and June Goodfield, *The Discovery of Time* (New York: Harper and Row, 1965), p. 165.

11. Ferris, *Coming of Age in the Milky Way*, p. 221.

12. John McPhee, *Basin and Range*, (New York: Farrar, Straus and Giroux, 1981).

13. Stephen Jay Gould, "Hutton's purpose," In *Hen's Teeth and Horse's Toes*, by Stephen Jay Gould (New York: W.W. Norton & Co., 1983), p. 86.

14. Stephen Jay Gould, *Time's Arrow, Time's Cycle: Myth and Metaphor in the Discovery of Geological Time* (Cambridge, MA: Harvard University Press, 1987), pp. 65–66.

15. Ibid., p. 78.

16. Gould, "Hutton's purpose," In *Hen's Teeth and Horse's Toes*, p. 86.

Chapter XIX. The Battle of the Geologists and the Physicists

1. Stephen Jay Gould, *Time's Arrow, Time's Cycle* (Cambridge, MA: Harvard University Press, 1987), pp. 257–258.

2. Charles Lyell, *Principles of Geology*, Vol. I, p. 123, as quoted in Gould, *Time's Arrow, Time's Cycle*, pp. 101–102.

3. Gould, *Time's Arrow, Time's Cycle*, p. 168.

4. G.J. Whitrow, *Time in History* (Oxford: Oxford University Press, 1988), p. 152.

5. Stephen Jay Gould, "The celestial mechanic and the earthly naturalist," In *Dinosaur in a Haystack: Reflections in Natural History*, by Stephen Jay Gould (New York: Harmony Books, 1995), pp. 32–33.

6. Immanuel Kant, *General History of Nature and Theory of the Heavens*, as quoted in Toulmin and Goodfield, *The Discovery of Time*, pp. 132–133.

7. For an excellent in-depth discussion of all of the early "clocks," see G. Brent Dalrymple, *The Age of the Earth*

(Stanford, CA: Stanford University Press, 1991), pp. 27–69. Material here is taken primarily from this source.

8. T.C. Chamberlain , as quoted in Dalrymple, *The Age of the Earth*, p. 44.
9. William Thomson, as quoted in Dalrymple, *The Age of the Earth*, p. 33.
10. Malcolm W. Browne, "New look at apocalypse: dying sun will boil seas and leave orbiting cinder," *The New York Times*, Tuesday, September 20, 1994, pp. C1 and C11.
11. James Marchant, *Alfred Russell Wallace: Letters and Reminiscences* (New York, 1916), pp. 205–206, as quoted in Loren Eisley, *Darwin's Century* (Garden City, NY: Doubleday, 1958), p. 240.

Chapter XX. The Clock That Worked

1. Susan Quinn, *Marie Curie: A Life* (New York: Addison–Wesley Publishing Co., 1995), p. 143.
2. Ibid., pp. 162–163.
3. Ibid., p. 163.
4. Lawrence Badash, "The age-of-the-earth debate," *Scientific American*, August 1989, p. 93.
5. A.S. Eve, *Rutherford* (New York: Macmillan, 1939), p. 107, as quoted in A. Hallam, *Great Geological Controversies* (New York: Oxford University Press, 1983), p. 101.
6. Quinn, *Marie Curie*, pp. 151–152.
7. Jacob Bronowski, *The Ascent of Man* (Boston: Little, Brown & Company, 1973), p. 344.
8. Quinn, *Marie Curie*, p. 173.
9. J.D. Macdougall, *A Short History of Planet Earth: Mountains, Mammals, Fire, and Ice* (New York: John Wiley & Sons, Inc. 1996), p. 86.
10. T.R. Dickson, *Introduction to Chemistry*, 2nd edition (New York: John Wiley & Sons, Inc., 1975), p. 48.
11. G. Brent Dalrymple, *The Age of the Earth* (Stanford, CA: Stanford University Press, 1991), pp. 87–88.

12. Arthur Holmes, *The Age of the Earth* (London: Harper and Brothers, 1913), p. 103.
13. Dalrymple, *The Age of the Earth*, p. 87.

Chapter XXI. The Past Recaptured

1. Nigel Calder, *Timescale: An Atlas of the Fourth Dimension* (New York: The Viking Press, 1983), p. 11.
2. G. Brent Dalrymple, *The Age of the Earth* (Stanford, CA: Stanford University Press, 1991), p. 75.
3. Lawrence Badash, "The age of the earth debate," *Scientific American*, August, 1989, p. 96.
4. Don L. Eicher, *Geologic Time* (Englewood Cliffs, NJ: Prentice–Hall, 1968), pp. 59–60.
5. Ibid., pp. 31–32.
6. Arthur Holmes, *The Age of the Earth* (London: Harper & Brothers, 1913), p. 165.
7. J.D. Macdougall, *A Short History of Planet Earth: Mountains, Mammals, Fire and Ice* (New York: John Wiley & Sons, Inc., 1996), pp. 18–19.
8. Claude Allègre, *From Stone to Star: A View of Modern Geology*, trans. Deborah Kurmes Van Dam (Cambridge, MA: Harvard University Press, 1992), p. 60.
9. Ibid., p. 63.
10. Mary Roach, "Meteorite hunters," *Discover*, 18:5, 1997, p. 73.
11. Dalrymple, *The Age of the Earth*, p. 401.
12. Ibid., p. 401.
13. Macdougall, *A Short History of Planet Earth*, p. 23.

Chapter XXII. Finding the Time of Humankind

1. Glyn Daniel, *A Short History of Archaeology* (London: Thames and Hudson Ltd., 1981), p. 35.
2. Glyn Daniel, *The Origins and Growth of Archaeology* (New York: Thomas Y. Crowell Co., 1967), p. 47.

3. Daniel, *A Short History of Archaeology*, p. 40.
4. Richard E. Leakey, introduction to *The Origin of Species*, by Charles Darwin (New York: Hill and Wang, 1979), p. 14.
5. Ibid., p. 14.
6. Alexander Marshack, *The Roots of Civilization* (Mt. Kisco, NY: Moyer Bell, Ltd., 1991), p. 64.
7. Joseph Prestwich, "On the occurrence of flint implements, associated with the remains of animals of extinct species in beds of a late geological period." *Proc. Royal Soc. London* 10, pp. 1860, 50–59, as quoted in Kevin Greene, *Archaeology, An Introduction* (Philadelphia: University of Pennsylvania Press, 1995), p. 14.
8. Rasmus Nyerup, *Oversyn over Faedrelandets Mindesmaerker fra Oldtiden*, 1806, as quoted in Daniel, *The Origins and Growth of Archaeology*, p. 80.
9. Richard Foster Flint, *The Earth and Its History* (New York: W.W. Norton & Co., 1973), pp. 47–48.
10. John Reader, *Missing Links: The Hunt for Earliest Man* (New York: Penguin, 1990), p. 150.
11. Don L. Eicher, *Geologic Time* (Englewood Cliffs, NJ: Prentice–Hall, Inc., 1968), p. 129.
12. C.C. Swisher III, W.J. Rink, S.C. Anton, H.P. Schwarcz, G.H. Curtis, A. Suprijo, Widiasmoro, "Latest *Homo erectus* of Java: potential contemporaneity with *Homo sapiens* in southeast Asia," *Science* 274, 1996, pp. 1870–1874; Stephen Jay Gould, "Unusual unity," *Natural History* 106:3, 1997, p. 20.
13. Richard Leakey, *The Origin of Humankind* (New York: Basic Books, 1994), pp. 4–8.
14. Ibid., pp. 8–9.
15. John A. J. Gowlett, *Ascent to Civilization: The Archaeology of Early Humans* (New York: McGraw–Hill, Inc., 1992), pp. 86–87.
16. Colin Tudge, *The Time Before History: 5 Million Years of Human Impact* (New York: Scribner, 1996), p. 37.

Chapter XXIII. The Clock That Came From Outer Space

1. Norman F. Smith, *Millions and Billions of Years Ago: Dating our Earth and Its Life* (New York: Franklin Watts, 1993), p. 92.
2. Don L. Eicher, *Geologic Time* (Englewood Cliffs, NJ: Prentice–Hall, 1968), p. 130.
3. John A.J. Gowlett, *Ascent to Civilization* (New York: McGraw–Hill, 1992), p. 198.
4. Stephen Jay Gould, "Up against a wall," *Natural History* 105:7, 1996, p. 16.
5. Nigel Calder, *Timescale* (New York: The Viking Press, 1983), p. 30.
6. Alexander Marshack, *The Roots of Civilization* (Mt. Kisco, NY: Moyer Bell, 1991).

Index

Page numbers in *italics* refer to illustrations.